Doing Good with Technologies

Philosophy of Engineering and Technology

VOLUME 4

For further volumes:
http://www.springer.com/series/8657

Katinka Waelbers

Doing Good with Technologies

Taking Responsibility for the Social Role of Emerging Technologies

Springer

Katinka Waelbers
Department of Philosophy
University of Maastricht
Grote Gracht 90–92
6211 SZ, Maastricht
The Netherlands
katinka.waelbers@maastrichtuniversity.nl

ISSN 1879-7202 e-ISSN 1879-7210
ISBN 978-94-007-1639-1 e-ISBN 978-94-007-1640-7
DOI 10.1007/978-94-007-1640-7
Springer Dordrecht Heidelberg London New York

Library of Congress Control Number: 2011930673

Printed on acid-free paper

Springer is part of Springer Science+Business Media (www.springer.com)

Contents

Chapter 1
Responsibility for the Social Role of Technologies

Technologies influence the way people act: when listening to an mp3 player or a mobile phone, people are less inclined to talk to for instance shop keepers at stations, and behave more "autistic" when using public transport. Internet and mobile phones have deeply influenced the way people communicate with each other and how they define, begin and maintain friendships. Some of these influences are desirable: people of European countries are more involved with wars, starvations and natural disasters that take place in other, less rich parts of the world, just by watching television. Consequently, they are more inclined to provide (financial) aid. But watching too much television is also to blame for the lack of sufficient exercise, and as a consequence many people in the west become obese by eating snacks while watching television.

To prevent undesirable social influences, can engineers and scientists try to develop technologies in such a way that their influence on people's actions is more positive? In road design, this idea is already commonly adopted: speed bumps, for instance, are constructed to encourage drivers to slow down. Why not apply the understanding that technologies influence people's actions to design a better world? This study aims to develop some ethical foundations and tools to enable practitioners—engineers, scientists and commissioners—to take a forward looking responsibility for the changes their technologies can bring about. This chapter begins with analyzing the presumptions behind this suggestion in more detail and provides an overview of the structure of the main argument of this book.

1.1 Designing Mediating Technologies

Consciously and unconsciously, engineers and scientists increasingly develop and employ technologies in such a manner that they steer, alter or influence our actions in what they regard as desirable directions. Examples of technologies that are intended to fulfill a mediating role are easy to find. Bruno Latour, for instance,

K. Waelbers, *Doing Good with Technologies: Taking Responsibility for the Social Role of Emerging Technologies*, Philosophy of Engineering and Technology 4,
DOI 10.1007/978-94-007-1640-7_1, © Springer Science+Business Media B.V. 2011

discussed how bleeps and flashing lights indicate that seat belts need to be fastened in such an irritating manner that it is almost impossible to not obey the technology. Latour also addressed tools such as the large key fob that encourages guests to leave their keys at the hotel desk, and the automatic door groom which prevents people from leaving the door open (Latour 1992).

When designing a technological artifact, people often have some implicit ideas about the practices of the eventual user and unconsciously develop mediating technologies (Akrich 1992). Consider, for instance, the length of the tube of the average vacuum cleaner: for many North European men, it is just too short for a comfortable working position and causes pain in the back. Such designs are inspired by traditional ideas on the division of roles between men and women and thus reinforce sexist assumptions about who should clean the house. A conscious redesign of such tools may stimulate men to also participate in cleaning tasks and therefore it may facilitate emancipation.

Since unconscious ideas about the eventual users can have undesired social effects, the Dutch philosopher Hans Achterhuis has put the idea of promoting "moralizing technologies" on the philosophical agenda (Achterhuis 1993, 1995, 1998). Exploring the full potential of Latour's sociological and philosophical work (Latour 1992), Achterhuis suggested that it is desirable to design technologies that stimulate us to drive safer, to adopt healthier lifestyles, and to behave in a more environmentally friendly way.

Internationally, the idea to design technologies for steering human actions has been put on the agenda by Jaap Jelsma under the title "behavior steering technologies" (Jelsma 2000). B.J. Fogg developed the concept of "persuasive technologies" for describing mediating computer technologies (Fogg 2003). Peter-Paul Verbeek makes a similar point when he argues for materializing morality (Verbeek 2006). In 2008 the influential book *Nudge* by Richard Thaler and Cass Sunstein—advisors of the United States President Barak Obama—put the idea of "choice architecture" on the international political agenda (Thaler and Sunstein 2008).

So, several authors support the increase of mediating technologies in our high-technological societies. But this is not uncontroversial. In the 1990s, Achterhuis' plea for moralizing technologies was received with two moral objections, the first of which was explained by Achterhuis himself (1998): (1) the social role of technologies is often more complex than foreseen, and undesired outcomes are common, and (2) the idea conflicts with the freedom of people (with their moral autonomy), as it subjects them to the rules of technologies (Swierstra 1999; Waelbers 2002). Further, it raises the questions of (3) who decides, based on which criteria, what actions are desirable? Let me shortly explain these three objections.

Since technological mediation is still not fully understood, the outcomes of technological mediation are often unintended. Some designs are meant to be—to use Alvin Weinberg's term—"a technological fix" for social problems (Weinberg 1966). Weinberg argued that many social problems could be better solved with engineering than by political, legal, or social means. Latour's examples seem to echo this suggestion on a very practical level. But Latour's work also makes clear that the so called technological mediation is unintended and even undesired: the social roles of technologies should not be oversimplified. Technologies are not simple means that can be employed for human ends without changing these ends (or "programs of action" as Latour prefers to call it). Latour argues that technologies affect or co-shape our actions, which can lead to substantial, unintended effects. The technical term he employs for the social role of technologies is mediation (Latour 1995a). Unintended outcomes occur because mediation is a non-linear process (Latour 2002). To put it in Latourian terms, by employing technologies, people take a detour and will not reach their initial aims (Latour 2002, p. 252):

> If we fail to recognize how much the use of a technique, however simple, has displaced, translated, modified, or inflected the initial intention, it is simply because we have changed the end in changing the means, and because, through a slipping of the will, we have begun to wish something quite else from what we at first desired. If you want to keep your intentions straight, your plans inflexible, your programmes of action rigid, then do not pass through any form of technological life. The detour will translate, will betray, your most imperious desires.

The results of technological mediation often turn out to be quite the opposite of what was intended. Edward Tenner calls this the "revenge effect of technologies" (Tenner 1997). A famous example is the first generation of energy saving light bulbs: it was meant to decrease energy consumption, but in effect, the technology achieved the opposite (Achterhuis 1998; Weegink 1996). Energy saving light bulbs are expensive to buy, but cheap to use. Consequently, many people started to use these bulbs for lighting places that used to be dark (such as gardens and corridors) and energy consumption increased. The outcomes of the human-technology interactions are often hard to imagine and, as a result, deliberately designing moralizing technologies is complicated.

The second objection is that the development of mediating technologies raises important questions concerning the freedom of our actions and thus concerning responsibility issues. Is it desirable for technologies to alter our actions or does this conflict with our autonomy? To what extent do they limit our freedom? These questions are of importance because if technologies direct our actions, it is questionable whether we can still be responsible. Reflection on such moral questions is scarce: commonly, practitioners and mainstream philosophers and sociologists ignore them. However, they are incredibly important because technologies are a substantial part of our lives and influence most aspects of our societies.

The idea of deliberately designing technologies to steer actions conflicts with the ideal of what it means to be a human being. According to many ethical theories, people's freedom should be promoted for it is necessary to exert their moral agency. The idea of delegating morality to a technical artifact is rejected by people who adopt a (neo-)Aristotelean ethics or a more Kantian ethics, because it leads to a loss of human virtuousness, moral intentionality, and moral freedom (Swierstra 1999). As Tsjalling Swierstra explains, from the viewpoint of virtue-ethics, an action that is externally initiated is a-moral, since the action is not the result of good character. From a Kantian perspective, a similar argument can be made since an action can only be called moral if the agent had the intention to act morally right. If he or she is seduced or forced to act, the action is not a moral action. With Kantianism, the mere reason to call an action moral is the good intention or the motivation that lay behind the action. So from both the virtue ethical as well as the Kantian perspective, one can argue that if technology is a co-steering force behind the action, the action becomes a-moral (Latour 1992; Swierstra 2000).

The most convincing response to these two objections is straightforward. Latour shows that technologies mediate our actions, regardless of whether we design them to do so or not. Achterhuis and Verbeek address our responsibility in this regard: if technologies are altering human actions anyway, it is better to design them in a manner that supports desirable actions (Achterhuis 1998). To put it in Verbeek's words (Verbeek 2006, p. 371): "when technologies are always influencing human actions, we had better try to give this influence a desirable form." All technologies are changing our actions. If people want to be free from such external force, they have only two options: refrain from using technologies altogether or take responsibility for the social role of technologies. Achterhuis and Verbeek suggest that we choose the latter: since technologies will be altering human actions anyway, we should take responsibility and design them in a way that makes us behave more "desirably". However, how can we take this responsibility if the human-technology interactions only become clear retrospectively, and when technologies are limiting our autonomy? And, there is still the third objection: who decides, based on which criteria what future mediating role is desirable? How can practitioners (such as scientists, engineers, and commissioners) prospectively explore and evaluate what the social role of the technologies will be? For such a task not only special forward looking tools are needed, but we also need some ethical foundations and democratic means to evaluate these social roles.

This study explores these objects for it aims to explore how practitioners can take a forward-looking responsibility for the future social role of technologies. It works to provide some ethical foundations and tools for broader reflection on the future technological mediation of human actions. Three research questions are abstracted from the objections:

1. How can the social role of technologies be best understood?
2. If technologies fulfill a social role, can people still be responsible?
3. How can practitioners work to take responsibility for the social role of future technologies in practice?

Ad 1. How can the social role of technologies be best understood? What do we mean when we discuss taking a forward looking responsibility for the social roles of technologies? First there is the question of what I mean by with technology. In philosophy of technology, there are four general usages of the concept of technology (Mitcham 1994): (1) technology as an object for referring to engineered artifacts such as cars, coffee machines, televisions, and the Internet; (2) technology as knowledge or a way of thinking for describing a particular manner of problem solving; (3) technology as volition for explaining that technology is also a psychological expression of our will to survive, control, and construct; (4) technology as an activity for addressing behavioral engagements with technological objects, knowledge or volition, such as the making of artifacts (engineering). Before the 1980s, philosophy of technology seldom addressed the first meaning. However, for studying how technologies and humans interact, we need to focus on the first meaning in which technology refers to a "thing". There are many differences between the distinct artifacts when it comes to mediation of actions.

Understanding the different forms of mediation is important for knowing the objects of the responsibility. Sometimes technologies have clearly definable risks, such as the risk of an airplane crashing or a gas tank exploding. But most human-technology interactions are less clearly describable and lead to more subtle outcomes, which cannot be expressed in probabilities. Technologies alter our choices, actions, and ideas, which can have substantial social and cultural consequences. Social and cultural changes, such as increasing individualization, are to a significant extent technologically mediated. Before discussing responsibility, we first need a clear idea about how we can best conceptualize these kinds of interactions: what mediating roles should we consider when thinking about taking a forward-looking responsibility?

Ad 2. If technologies fulfill a social role, can practitioners still be responsible? As pointed out above, technologies limit our freedom by influencing our actions. This is problematic since freedom is often seen as an essential condition for taking responsibility: you are not morally responsible if someone or something is forcing you to act. How much room is there to actually take responsibility and influence the courses of technological mediation? Some philosophers have argued strongly that technologies determine our society and that their development has an internal logic that does not allow for any adjustments from the human side (see Chap. 2). If they are right, it is unfair to ask people to accept responsibility for the social role of technologies. But as will be argued when analyzing the first question, the social role of technologies is not purely deterministic and there is still some room for exploring responsibility issues.

It is important to stress that this study adopts a different notion of responsibility to that common in ethics. Many philosophers focus either on liability or obligations. In the case of liability, being responsible refers to being the rightful target of responsive attitudes: you have done some right or wrong for which you ought to be praised or blamed. Obligation stresses that you are being responsible for something in the sense that it is your duty to do certain things. For the purposes of this study, the interest is not in these two kinds of responsibility, but is focused on responsibility for the social role of future technologies. The terms "praise" and "blame" are commonly used retrospectively, after the actions have taken place and when it has become clear what the consequences are. Duties can be assigned prospectively, but only if it is clear what a responsible action is. This is often unclear when discussing new technologies or new uses for technologies, and so duties cannot be defined. The notion of responsibility here is closely connected to Carl Mitcham's understanding that taking responsibility means carrying out a reflective analysis with "explicit consideration of ethical issues" (Mitcham 1997). Additionally, he argues that people involved in the development of technologies should have tools that are "sufficiently complex to include a diversity of non-standard technical factors" (p. 275). This study works to provide such tools for including a wider diversity of "non-standard factors" in the sense that it addresses a wider diversity of morally relevant factors than just economics, environmental issues and health risks. The aim is to draw attention to the interaction between technologies and the morally relevant actions of people. Taking responsibility in this broader sense implies questioning why certain aspects of the technology are good or not good from a certain moral point of view.

Ad 3. How can practitioners work to take on this responsibility for new technologies or new employments of technologies in practice? So far, the questions have addressed the social role of technologies and responsibility theoretically, but to be useful, people developing the technologies (the practitioners) need tools to assess individual technologies. Here we stumble upon the question of foreseeability, which is a philosophical and practical matter. For instance, one of the aims of introducing computers to the work floor of many businesses was to reduce paperwork, but both the work and the paper have increased vastly (Tenner 1997). This increase can only be explained by the interaction of new computing technologies (which facilitates easy storage of data and easy printing of documents) and the working habits of many people (who like to be optimally informed and who like to read from paper). As a result, significantly more paper is used. Additionally, computers help to change the way we think about information. Instead of reducing bureaucracy, computers have supported the introduction of many additional ways to collect, store and rework data. Both impacts were unintended and unexpected, and therefore, they were uncontrolled. For taking responsibility in such cases, practitioners need to be able to obtain some ideas about what the possible social role of the technology could be.

Here, "practitioners" is a technical term and addresses people involved in the actual development of the technologies such as engineers, scientists, and commissioners (individuals and collectives). This is not to argue that other actors such as users, politicians and citizens do not have any responsibility of their own, but they play a different role in the development and design of technologies and so taking a forward looking responsibility for them will involve other steps, which this study does not address. The focus in this study is on enabling practitioners who are professionally involved in the processes to take a proactive responsibility by developing pragmatic tools for exploring and evaluating the possible social roles of their technologies.

In dealing with these three questions, this study makes use of Bruno Latour's and Alasdair MacIntyre's work. Or, to put it more accurately, this study aims to combine Latour's insights on technological mediation with MacIntyre's ethical and anthropological ideas. The French sociologist and philosopher Bruno Latour (Beaune, June 22, 1947) has studied the techno-social construction of science and is founder of the influential Actor Network Theory which explains how to study the interaction between humans and nonhumans (technologies). The Scottish philosopher MacIntyre (Glasgow, January 12, 1929) spent his successful career reviving an Aristotelian moral philosophy and anthropology that is based on the study of traditions, practices, and virtues. These short descriptions may reveal the difficulties of trying to develop ethical foundations and tools based on the work of both authors at once: they differ in both object of study and methodology. Nevertheless, I argue that combining some aspects of their work is fruitful for the aim of this study. The main structure of the argument is outlined in the next section.

1.2 The Main Argument of This Study

This section introduces how the main argument of this study works to respond to the three questions.

Chapter 2 To understand the social role of technologies, Chap. 2 explores the main positions taken in the twentieth century in philosophy of technology and the mainly sociological field of Science Technology and Society (STS) studies. These fields often work together to develop a broader understanding of the complexity of the human-technology interaction. In this regard, one of the most influential schools in STS is based on Latour's theoretical and empirical work (Latour 1991, 2005a). Although the issue of the subtle but substantive ways in which technologies influence human actions has long been studied, by—among others—historians (Mumford 1934), philosophers (Anders 1980/1956; Heidegger 1962a), sociologists (Ellul 1964), social critics (Illich 1973), and political theorists (Winner 1980), Latour has restated many points in ways that have become the focus of various academic debates. Latour has studied philosophy and consequently his work is sociological as well as philosophical in nature. He writes: "even though I have always held positions in sociology, and have sometimes been accepted as an honorary anthropologist, and feel much loyalty to the little field of science and technology studies, and have also dabbled in social theory, I have never left the quest for philosophy" (Latour 2008, p. 2).

Latour's approach is helpful for understanding the social roles of technologies, and it also provides a usable answer for the pessimistic philosophers who claim that technologies are determining our societies and our actions. Actor Network Theory (ANT) teaches that even though the human technology-relation is particularly complex, it can be made intelligible by addressing the mediating role of technologies. But, mediation theory also makes it clear that the atomist rational human agent that is presumed by dominant moral theories such as deontology and utilitarianism is a fairytale. We are not free to determine our own ends and means: ANT stresses that other humans and technologies are continuously mediating our actions. This understanding raises serious problems for the possibility of taking responsibility.

Chapter 3 Chapter 3 explores the problems ANT raises for responsibility issues. These problems arise because of the strong emphasis on the similarities between the actions of humans and technologies. Since both fulfill comparable social roles, humans and technologies are both called agents (or actants, which is the same in ANT) who only exercise agency together. In other words, agency is the result of the interaction between human and technological agents. This is not simply an innocent provocative way of saying that technologies mediate human action in ways that are comparable to how people mediate each other: it has deep, controversial consequences (see also Waelbers 2009b).

First, the vocabulary of ANT does not offer much space for discussing the rudiments of human moral agency. ANT stimulates reflection on desired and undesired technological impacts, but it does so by using an output and power oriented understanding of agency. However, the concept of moral agency is commonly used in moral philosophy to draw attention to a wider palette of human motives—which also includes desires, intentions and moral beliefs—and to the human capability to reason about the desirability of their actions (that is, to take responsibility). If we want to take responsibility, we need to understand ourselves as moral agents, distinct from will-less objects.

Second, ANT describes all interactions between humans and technologies in terms of power relations without offering a stratified understanding of how technologies alter our actions. Within the vocabulary of ANT, one can proclaim that the microwave interacted with the family dinners, genomics research interacted with our understanding of humanity, and speed bumps interacted with the behavior of automobile drivers. Technologies fulfill interacting roles, meaning that they "determine", "backdrop", "authorize, allow, afford, encourage, permit, suggest, influence, block, render possible, forbid, and so on" human action (Latour 2005a, p. 72). If an idea can be applied that broadly, the question of what it exactly means becomes pressing. If the aim is to take responsibility for the social role of technology, we need tools to understand the mechanisms of interaction on a more detailed level.

Third, one can argue that we can only take responsibility for new technologies or new employments of technologies if we can "open-up the black-box" prospectively. Acting knowingly becomes more difficult when accepting ANT, since it teaches that the impacts of technologies are not limited to risks and clear definable side effects such as environmental consequences and biomedical harm. Latour shows that technologies have unexpected and undesired social impacts because they alter our actions leading to entirely new situations and impacts. This process is described by Latour as a "black-box" (Latour 1994), and allows few forward-looking possibilities.

These three problems call for responses:

- We need an understanding of human agency acknowledging that we are part of larger socio-technological networks, and that we are moral agents with multiple motives;
- We need a more layered concept of mediation that tells something about how technologies alter human action, and;
- We need some practical tools for imagining the social role of technologies.

Chapter 4 To be able to recognize the importance of Latour's work, while simultaneously making room for ethical reflection, requires a moral philosophy receptive to the idea that people act and live in a social context which provides them with reasons or motives for actions. It is important to see, as is often wrongly presumed, that humans are not atomistic subjects who can freely determine their own course of action.

Chapter 4 explores whether such a moral philosophy can be found in the practice oriented ethics of Alasdair MacIntyre. MacIntyre elaborates extensively on what it means to be a human agent situated in a social setting (MacIntyre 1985). MacIntyre's concept of human agency looks promising for a context oriented understanding of responsibility, since he combines human social dependence and biological condition with belief in the importance of human reasoning (MacIntyre 1999). He argues that reasons for actions (motives or presumptions that an action contributes to and aimed for good) are molded biologically and socially (and so, we are not autonomous, Kantian individuals), but still MacIntyre maintains we can regain some autonomy by using our capacity of practical reasoning—the capacity of pragmatic reflection which people exert on a daily basis when they ask themselves what is best to do—to become conscious of our reasons and to assess them.

However, MacIntyre never discussed ANT, nor has he elaborated in his writings on questions regarding technology. Therefore, it has to be examined carefully which elements of his philosophy can be adopted to enrich ANT, and which elements of ANT are or are not reconcilable with MacIntyre's moral philosophy. An important difference between the two approaches is that Latour adopts an output oriented (focusing on the consequences of action) and social power directed account of agency, while MacIntyre focuses on attitudes and intentions of action and has made clear his aversion to the power discourse of authors such as Friedrich Nietzsche and Michel Foucault (MacIntyre 1990).

Chapter 5 When it is clear to what extent MacIntyre's concept of moral agency can be employed to develop a practical ethics reconciled with the empirical findings of ANT, we can start working in Chap. 5 to reformulate a mediation theory that simultaneously respects differences and similarities between the social roles of humans and technologies. Latour's mediation theory explains that technologies change actions, but this understanding is too general. Mediation theory needs to be reformulated with a vocabulary that provides the necessary space for taking responsibility.

Chapter 5 searches for a formulation of mediation theory that, on the one hand, provides a deeper understanding on the mechanisms of technological mediation, while, on the other, incorporates the MacIntyrean idea of moral agency. The key notion is that technologies do not merely mediate our actions, but they mediate our motives or reasons for action. We can take responsibility insofar as we are capable of using our capacity of practical reasoning to reflect on the mediation of these reasons.

In this way, mediation theory becomes less controversial in the sense that technologies are not seen as agents. However, if we take seriously the idea that technologies alter our reasons for actions, we also have to reconsider the implications of technological mediation on a more radical level. MacIntyre's ethics teaches that our moral beliefs—one of the most important reasons for our actions—are largely practice and tradition based. This can change over time due to technological mediation. Chapter 5 argues—inspired by the ethics for New and Emerging Science and Technology (NEST-ethics) (Swierstra and Rip 2007) and by post-phenomenology

(Verbeek 2005b)—that technologies mediate various reasons for actions, including moral beliefs.

Chapter 6 With the viewpoint developed in Chaps. 1–5, Chap. 6 turns to the question of how to take responsibility for the social role of technologies. Human agency has reacquired a possibility space for some sort of autonomy, without ignoring the biological, social, and material contexts. Mediation theory has been reformulated to open up some space in which responsibility can be taken. Based on these notions, it is appropriate to elaborate on how we can actually take responsibility when adopting the extended mediation theory in which our reasons for actions are mediated. The renewed concept of technological mediation promotes imagining possible broad impacts of technologies.

Chapter 6 draws our attention to this aspect and aims to develop a preliminary toolbox for taking responsibility for the social role of technologies. The toolbox consists of three parts:

- A five-step approach to structure reflection about the possible social roles of a technology;
- An introduction to some existing academic approaches that study possible future human-technology interactions, and;
- Several common sense criteria for evaluating the results of the enquiry that are based on the premise that technologies should support human and environmental flourishing.

Chapter 7 Chapter 7 shows how these tools for taking responsibility can be applied in a realistic case of road safety technologies. This case study is based on the European Commission's (EC) promotion campaign for the "car of the future". The question is whether the positive view of the EC will become true. Will the cars of the future be safer, environmental friendlier and prevent congestion, or should we fear "revenge effects"? Additionally, how does the car of the future alter beliefs about desirable driving styles and responsibility issues? How might it influence our experience of driving a car? These questions are discussed in a structured manner, with the guidance of the five-step model.

Chapter 8 The conclusions of Chap. 8 explore whether practitioners can take responsibility for the future social role of technology. Does the elaborate discussion of the three questions in the previous chapters indeed provide sufficient response and the tools to take responsibility in practice, or are there other substantial obstacles that still hinder the acceptance of responsibility?

Chapter 2
Mastering Technologies

To enable practitioners to take responsibility for the social role of technologies, it first has to be clear what this social role actually is, and whether it can be influenced by engineers and other professionals involved in technology development. Since Friedrich Engels and Karl Marx, multiple philosophers and sociologists have studied the social role of technology. The first part of this chapter characterizes the positions of several of these authors by explaining their answers to two questions: (a) how deterministic is the technological influence on society? and (b) who or what determines the technological development? The second part of this chapter turns to Latour's approach to see how his view may help to overcome the seemingly contradicting answers that the authors gave to these questions.

2.1 Traditional Debates on the Social Role of Technologies

Since it is only fair to ask people to accept responsibility for something that is well understood and within their control, it is important to obtain a clearer view of the social role of technologies. In other words, for taking responsibility, we need an accurate understanding of the human-technology interaction: what are the effects and can we influence them? Traditionally, it was commonly argued that technological change is autonomous, which implied one of both of two things (Mitcham 1985, p. 97):

> One, an internal autonomy born of the necessary evolutionary sequence in technical forms; two, an external autonomy concerned with the inevitability of certain societal effects. In other words, technological change can be autonomous in the sense that it evolves according to its own internal principles; and/or it can be autonomous as the more determining than determined factor in society—determining, that is, social relationships, culture, etc.

Though it is frequently found that authors defend both forms of autonomous technological change, the positions are also defended separately: referring to the internal logic of technological development is different from referring to how

K. Waelbers, *Doing Good with Technologies: Taking Responsibility for the Social Role of Emerging Technologies*, Philosophy of Engineering and Technology 4,
DOI 10.1007/978-94-007-1640-7_2, © Springer Science+Business Media B.V. 2011

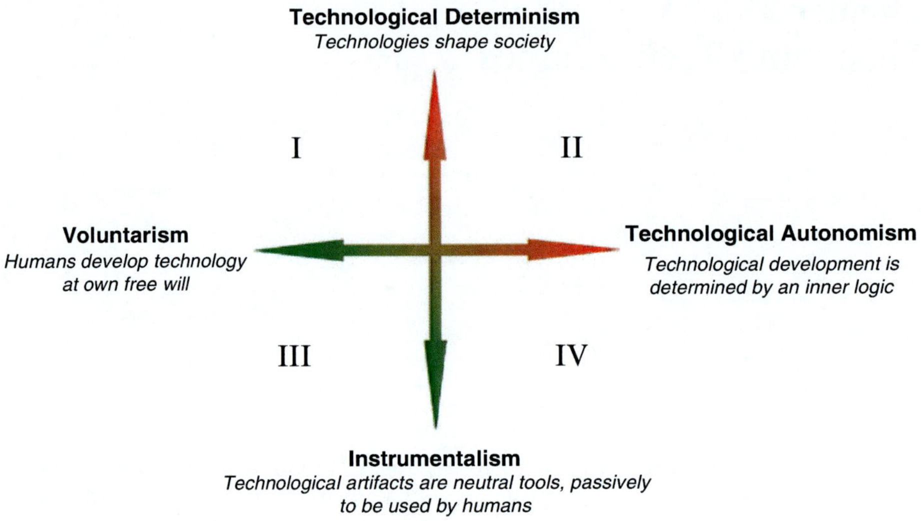

Fig. 2.1 The two spectra presented schematically

technologies shape the social. This line of thought is used in this chapter to structure the philosophy of technology debates of the first three quarters of the twentieth century in order to help understand different interpretations of the human-technology interaction. The mainstream philosophy in this field of study altered substantially in the late 1980s due to the empirical turn (Meijers 2000). Traditional philosophy of technology refers to debate during the first three quarters of the twentieth century. This change is to a large extend inspired by Latour, whose theory will be discussed at lengths in the second half of this chapter, as well as in the upcoming chapters.

A matrix consisting of two axes (see Fig. 2.1) that results from distinguishing developing technologies from adopting technologies, can help identify the most important lessons to be learned from the early philosophy of technology debates. The first axis represents a spectrum ranging from belief that technologies are fully shaped by humans to belief that technological development is an autonomous process (see also, Mitcham and Waelbers 2009). On the one extreme you find *voluntarism* stating that people design the technologies and their functions according to their own free will. Voluntarists believe that people create technologies freely and argue that technological development is fully malleable. Consequently, people can take full responsibility for the shaping of technologies. On the other extreme, *technological autonomism* is placed. Technological autonomists describe technological development as an autonomous processes, fixed by an internal logic. According to this view, one technological development necessarily follows the previous, based on natural laws and laws of, for instance, efficiency. If people embrace technology, they have to accept responsibility for this embracement, but they cannot accept responsibility for the actual manners in which technologies develop. In other words, they can choose for or against technology, but they cannot choose the shape of technology.

The second axis represents a spectrum of positions concerning the social role of technologies. This spectrum opposes *technological determinism* to *instrumentalism.* Technological determinism sees technologies as a fundamental force behind cultural and historical change, and they understand contemporary human culture as a kind of "technopolis" (Postman 1992). According to the determinists contemporary society is completely shaped by technology. Therefore, the object of responsibility in this discussion is extremely wide: it refers to the complete technological shaping of our social order. Of course, it is unrealistic to accept a proactive responsibility for something which is so all-encompassing. Instrumentalists reject such analysis and argue that technologies are simply tools that only influence societies because we want them to. They claim that people shape their own cultures, for which technologies are used at free will. This view entails that technologies do not influence our actions substantially and therefore we should accept responsibility for the use or abuse of a technology.

These two spectra can be plotted against each other in a matrix representing four different positions. Two positions of this matrix are most common, since determinism often implies some form of autonomism (quadrant II), and voluntarism is often combined with instrumentalism (quadrant III).

Jacques Ellul is a strong representative of quadrant II, since he argues that in technological development the most efficient technology is automatically chosen, and that technological changes deeply shape all elements of contemporary societies (Ellul 1964, 1989). Per Sundström (1998), Samual Florman (1976) and most engineers defend the opposite position, and thus represent quadrant III. They accept a combination of voluntarism and instrumentalism: technologies are designed following human desires and the tools are freely used.

Most scholars are found in quadrants II and III, but representatives of quadrants I and IV can be found. Influential representatives of quadrant I are David Collingridge (1980) and Hans Jonas (1984). Jonas, for example, holds the position "that with certain [free] developments of our powers the nature of human action has [necessarily] changed". This change "calls for a change in ethics as well" (Jonas 1984, p. 1). Karl Jaspers (1957, 1962), the Club of Rome (Meadows et al. 1972) and Nick Bostrom (2005a) are well-known representatives of quadrant IV. Again, their interpretations differ, but Bostrom argues for instance that we should take advantage of the unavoidable technological developments to become better creatures.

For learning more about the social role of technologies, it is useful to consider some of these thinkers in more detail, and thus examine more fully the four positions.

2.1.1 Technological Autonomism Versus Voluntarism

One of the most influential authors who described the belief in "*autonomy of technique*" was Jacques Ellul (1964). Ellul argued that in our pro-technological world new superior techniques will always push aside the less advanced ones and that societies are compelled to adjust itself. He describes this phenomenon as "the quest of the one best means in every field" (Ellul 1964, p. 21). In contemporary society, many people believe there is only one best way in technology, Ellul argues, and this best way is fixed by the inner logic of technological development (see also Mitcham 1994). With their urge for efficiency, people prescribe the best way only economically (efficiency) and therefore determine which techniques survive. Ellul defines technique as "the totality of methods rationally arrived at and having absolute efficiency (for a given stage of development)" (Ellul 1964, p. 25). Since technique is a pre-eminent for increasing efficiency, a lack of efficiency is, according to Ellul, seen as a technological problem that requires a technological solution.

As a result, the technological system enforces itself. Ellul calls this "self-augmentation". New techniques come with problems that are solved with newer techniques which, in turn, bring new problems that call forth technological solutions.

Examples are not hard to find: due to technological development, we need electric power plants. Traditional power plants use non-renewable sources of energy such as coal and oil, which cause air pollution. For solving this problem a technological solution was created in the form of nuclear power plants. However, this solution brings other serious problems such as nuclear waste and risks of accidents. Nowadays engineers and scientists are working hard to find technological solutions for these new complications.

Ellul argues that due to the pro-technology attitude of most people, technological development has become autonomous in multiple ways. Technological development is—among other things—independent from economics and politics. Neither economic nor political evolution conditions technical progress. Technique is also autonomous with respect to morality and spiritual values in the sense that external critique is not tolerated. Putting limits to technique based on values is different for people, according to Ellul, because technique presents itself (misleadingly) as value-neutral. Since it is based on mathematical, physical and biological laws, the hidden values are not easily recognized. These laws are the foundation on which technique functions, together with people's pursuit of efficiency. Because technique must gain its results through mathematical precision, there is no room for human variability. Therefore, it replaces human beings and the work they do (Ellul 1964, p. 146):

> But this delusion [the triumph of humans because of technique] cannot last much longer. The individual obeys and no longer has victory which is his own. He cannot have access even to his apparent triumphs except by becoming himself the object of technique and the offspring of mating of man and machine.

The relation between techniques and technologies is complex in Ellul's work. "Technique" does not refer to just technology, technological artifacts, or machines. Technique is a methodological framework of structuring the world. While people are striving for efficiency, technology turns everything into means. Technique begins with machines (apparatus) and without machines the world of technique would not exist (Ellul 1964). So technique includes technological devices, but it also surpasses them. Nonetheless, within Ellul's work, technological devices are technique per se.

However, in the last third of the twentieth century, sociological Science and Technology Studies (STS) started to agree that the autonomist view was too limited, as well as too negative. Although, engineers are not free to create a perpetual motion machine, efficiency and striving for more advanced technology are not the sole factors that determine technological developments. Consider for instance iPods: style and simplicity were the main ideas for the development of these mp3 players. The concept of technique turning everything into a means for the sake of efficiency is too limited. More aspects seem to be essential.

When analyzing a technological development retrospectively, it may seem that it was merely efficiency-driven, and that other values did not play a role. But that is only the result of practicing "Whig history" (Sismondo 2004). Linear perceptions of history based on efficiency arise because people attempt to "construct the past as a series of steps toward (and occasionally away from) present views" (Sismondo 2004, p. 18). Many scholars defending technological autonomy implicitly presume a teleological view that developments in the past lead directly to the current state of affairs as if there were a strict line of progress in the course of events. Technological development is not that straightforward: multiple other values (such as economic, political, and aesthetical values) also play important roles.

To support this point, Wiebe Bijker developed the theory of the Social Construction of Technology (SCOT), for example (Bijker 1984; Bijker and Law 1992; Bijker et al. 1987). SCOT is a direct reaction to the story of technological autonomy and offers an influential method for empirical social study of technological development that presumes (conditional) voluntarism. SCOT teaches that technologies do not succeed because they are more efficient, but they are seen as efficient because they are accepted or appreciated by humans: people give shape to technological change, and therefore they can take responsibility for this change.

The SCOT approach is frequently used—sometimes with minor changes—to conduct empirical social studies, for example (Kline and Pinch 1996; Rosen 2002). Central to this theory is that (Rip et al. 1995, p. 3): "social effects of any technology depend crucially on the way impacts are actively sought or avoided by actors, involved in the development of technology". Actors—companies, non-profit organizations, engineers, users and governmental organizations—can successfully exert power, which results in developing certain technologies *and* the denial of others. For example, though we are technologically capable of cloning humans, most states and research institutes forbid developing this technology for moral reasons of for fear of public reaction. Freedom, deliberation and democratic power structures are often regarded as conditions for the ability to influence developments (Rip et al. 1995), as well as participating in the early stages of the developmental process (Collingridge 1980).

The basic claim of voluntarists is that technological development is a social process, meaning that within the limits of the laws of nature, humans freely create what technologies they want. Sundström has formulated a definition of technology that suits this view: "technology is an application of a scientific, mathematized and value-free (or value-neutral) account of nature, from which it inherits, as it were, a programmatic methodological value-neutrality" (Sundström 1998, p. 44).

As an aside, it cannot be denied that there is a relation between the development of scientific knowledge and methodology on the one hand and technological development on the other. But Sundström seems to assume a linear relation between natural science and technological development. Technological developments with little scientific foundation—as often found in history, for instance Medieval cathedral building and Roman crossbows—are not taken into account. Furthermore, the assumption states that we translate natural science into technology, without the introduction of other cognitive elements such as engineering knowledge and skills, designers' values and tastes. Technical knowledge and methodology differ from scientific knowledge and methodology. Of course, scientific concepts can play an important role in technical development, but scientific concepts alone are not enough. In order to develop a technology, engineering theory, technical skills and more diffuse data—such as common knowledge and practice-related information—are of great importance (Staudenmaier 1985).

Sudström assumes that the natural sciences (or at least the methodologies of the natural sciences) and the practice of engineering are endlessly malleable and value-free. He argues that we are able to interpret and reshape the world around us with hardly any restrictions.

However, other empirical studies have shown that technological development is not that free (Smits and Leyten 1991). The QWERTY keyboard was deliberately developed to prevent people from typing too fast. In traditional typing machines, the hammerheads got stuck if someone typed too fast. In contemporary equipment, typing too fast is no longer a problem, yet we still use the same keyboard. Several attempts have been undertaken to introduce new, more efficient and ergonomically improved keyboards, but mainly due to the social structures in which the technology is deeply embedded, these attempts failed in most countries. In other words, once chosen for a certain type of technology, a path dependency aries that gives people the idea that there was no choice at all. For example, secretarial colleges are tailored to the QWERTY system and most people do not want to relearn typing (Smits and Leyten 1991). It is hard for individuals to choose for another system. Engineers depend on already implemented standards and many actors, such as competitors, governmental organizations, suppliers of materials, who work with them (Smits 2000).

Summarizing, the internal dynamics of efficiency and internal drive to choose for the technologically more advanced approach are not the only driving forces in technological development. To obtain a better understanding of whether and how we can influence technological change we need a position that enables us to combine the valuable insights of both positions.

2.1.2 Instrumentalism Versus Technological Determinism

The discussion between instrumentalism and technological determinism resembles the discussion between voluntarism and autonomism. The reason is that autonomism is often intertwined with determinism. When Ellul defines technique as a framework that turns everything into a means for the sake of efficiency, the driving force of technological development (efficiency) is closely connected with the broad impacts of technologies Ellul addresses: technology is turning everything into means. But denying autonomism does not automatically lead to the dismissal of determinism. Collingridge argues that, at least early in the process, technological development is free. If it is accepted that efficiency is not the only driving force, and that other objectives can be incorporated in the technological system, choices can be made between several technological alternatives. However, this does not imply that technologies do not modify society. As Collingridge explains, once the process is more advanced, the choices made earlier become more rigid. Altering or steering the technologies becomes more difficult as the social and technological embedding becomes fine-tuned to the chosen technological course. As a result, when the technology is implemented in society, people often have to adapt their habits or actions in order to ensure that the technology functions, as it is too late to change the invention (Collingridge 1980). Hence, a "soft version" of determinism is conceivable without autonomism.

Determinists stress that the state of technology is the most important author of social structures, because technology enables human action (Sismondo 2004). As Ellul says (Ellul 1964, pp. 5–6):

> Technique integrated the machine into society. It constructs the kind of world the machine needs and introduces order where the incoherent banging of machinery heaped up ruins. It clarifies, arranges and rationalizes; it does in the domain of the abstract what efficiency does to everything. Moreover, technique is sparing in the use of the machine, which has traditionally been exploited to conceal defects of organization.

The idea that machines structure society, and therefore make history, leads the determinist's approach (Misa 1988). They argue that social developments such as urbanization should be seen as the a result of the industrial revolution. Karl Marx and Friedrich Engels argue that there is a clear relation between technological developments on the one hand, and the historical development of economic systems and power relations on the other. As Marx said, "the hand-mill gives you society with the feudal lord, the steam-mill, society with the industrial capitalist" (Marx 1963/1847, p. 92).

Determinism arose in the context of the Industrial Revolution, two World Wars, and technologically induced social and cultural change. First, there was the industrialization of production. Efficiency of production was the main argument for implementing assembly lines in the factories, forcing people to work mechanically and to focus mainly on the function of the produced goods and less on the aesthetical aspects. These developments led to technological determinism and alienation thesis (Verbeek 2005b). Second, in the World Wars many new technologies for increasing battle effectiveness turned out to have devastating effects such as trench warfare, the spread of poisonous gasses and the nuclear bombing of an island. Apocalyptic questions concerning a possible third World War were raised and the idea of human subordination was added to the philosophers' agenda by (among others) Günther Anders (Anders 1980/1956) and Arnold Gehlen (Gehlen 1961).

Third, due to strong technological development, life has changed dramatically: technologies such as electric lighting, telephones, automobiles and planes were all introduced within the lifetime of one generation. Technologies reshaped life substantially and many authors had the feeling that culture and everyday life were completely determined by technology. Postman, a late representative of this techno-pessimism, introduced the concept of "technopolis" (Postman 1992) to address the all-encompassing social and cultural impacts of technology.

In the first three quarters of the twentieth century, sociologists and philosophers generally adopted pessimistic versions of technological determinism. But the commonly more optimistic instrumentalists argued against this world-view as it ignores the human influence. Instrumentalism is often found among scientists and engineers, and they have two strong common sense arguments for their claims. First, they defend the principle of ambivalent use. This principle states that technologies are value-neutral, because in themselves they are not good or evil. People use them for better or worse purposes, but that is not a characteristic of the instrument itself. Ambivalent use—or double use (Briggle 2005)—means that most techniques or tools can be used in different ways, or for different purposes, each with its own positive or negative value characteristics (Sundström 1998). This holds—according to Sundström—as long as the technology has not been designed or made for a purpose that restricts the possibilities of use, such as an atomic bomb. In this sense, technologies do not determine human action. Humans can use most technologies for many different actions, as is often succinctly illustrated with the example of the knife: one can use it to peel and cut an apple or to murder one's spouse.

But does ambivalence mean that technology is value-free or non-normative? As Ellul (Ellul 1964) has shown, technologies are themselves ambivalent. Often, it is no longer possible to choose between good and bad use of contemporary complex technologies, as in the case of traditional, simple technologies such as knives. Cars, for instance, bring good and bad things at the same time. People have become more mobile, which is often regarded as being valuable. But on the other hand, it also brings casualties and environmental pollution. Additionally, cars demand a change of our landscapes. This wide variety of impacts cannot be exclusively explained by dual use: they are brought about by the intended use.

Second, instrumentalists explain the principle of inaction (Sundström 1998), which means that when not in use, the technology does not result in anything good or bad. If a knife is not used, it does not bring any harm either. Technologies have no value properties as long as they are not used or applied in practice (Sundström 1998). Only when they are used by humans—who are value-imparting beings—is a technology value-loaded. So, as long as humans do not use a given technology, it remains value-neutral, according to the instrumentalists.

However, it can be questioned whether it is true that only the *use* of a technology has impact on the morality of a situation. Also the mere availability of a technology is enough to bring about change. Not using your car for short distances has a direct positive effect on air pollution, but town and country planning will not be affected by this decision. Furthermore, if one can choose to use a technology, the non-use of the technology becomes an act as well. For example, choosing to use or not use prenatal screening for Down-syndrome implies reflection on (moral) questions such as abortion, malleability of life, and quality of life. The mere availability of the test changes the whole morality of pregnancy. It is no longer "fate and destiny" but human conduct that determines the future of the fetus. This is even true when the woman decides not to use the test (Verbeek 2008b).

Additionally, social and psychological aspects make it difficult to say "No" to technologies. According to determinists, this is even impossible (Ellul 1989, p. 23): "technique is the complex and complete milieu in which human beings *must* live, and in relation to which they *must* define themselves". Think about the increasing pressure for everyone to use internet, uphold several e-mail accounts and own a mobile phone. If you want to engage in society (with or without a paid occupation), you need to embrace these technologies.

Does this all imply that the determinists are right? Alongside the already mentioned examples of the keyboard and human cloning, there are many other examples of technologies that failed to become a success. Social studies on the development of telecommunication have shown that the role of the users is substantial. When the first telephone exchange was established in the USA in 1878, it was only intended to be used for business purposes. It was seen as too expensive to use for chatting, and it was regarded as indecent to employ such a high tech system for such frivolous means. Private conversations were simply cut short by the switchboard operator. However, the demand for using the system for "just" private matters and chatting was also great, and finally, the system was adapted for this use. Furthermore, it was seen as lazy to make a phone-call to someone within walking distance of half an hour, but eventually people adopted the technology anyway for all kinds of social uses (De Wit 1998; Smit and van Oost 1999).

I have structured the traditional philosophy of technology debates with the help of the matrix to show that all four positions have valuable lessons to teach us. But in doing so, it also becomes clear that none of the four positions actually holds for they are mutually exclusive. In the 1980s however, traditional philosophy of technology changed dramatically. An empirical turn in the field was enforced in the attempt to study the social role of specific technologies in distinct social contexts instead of trying to describe the human-technology interaction in more overarching terms.

2.2 The Empirical Turn in Philosophy of Technology

One of the first philosophers to make this step was Langdon Winner. His starting point is to stop thinking about technology as a system of thought or a structuring principle, and to stop generalizing individual technologies as simple tools. Instead, by combining the work of multiple disciplines studying the social role of technologies, Winner shows that individual artifacts can play a political role (Winner 1988, p. 123):

> One strength of this point of view is that it takes technical artifacts seriously. Rather than insist that we immediately reduce everything to the interplay of social forces, it suggests that we pay attention to the characteristics of technical objects and the meaning of those characteristics.

Winner aims to go beyond descriptions of *making* and *using* to "examine critically the nature and significance of artificial aids to human activity" (Winner 1988, p. 749). There is more to technological change than commonly recognized in the ways inventors, engineers, operators and repair technicians make and maintain artifacts that others can pick up, use, or set aside. For Winner both voluntarism and instrumentalism constitute "technological somnambulism" in which we sleepwalk and fail to recognize the extent to which technologies reshape human activity and its meanings. Adopting the concept that Ludwig Wittgenstein uses for language, Winner argues that automobiles, electric lights, and computers have become "forms of life" that create a culture that is scarcely conceivable without them. Reflecting on the role of television, winner writes (Winner 1988, p. 257):

> None of those who worked to perfect the technology of television in its early years and few of those who brought television sets into their homes ever intended the device to be employed as the universal babysitter. That, however, has become one of television's most common functions in the modern home. Similarly, if anyone in the 1930s had predicted people would eventually be watching seven hours of television each day, the forecast would have been laughed away as absurd. But recent surveys point out that we Americans do spend that much time, roughly one-third of our lives, staring at the tube.

But Winner is also critical of autonomism and determinism. Watching television is a choice, even though turning it off is not as easy as instrumentalists customarily assert. TV is woven into the fabric of daily life with programs that are topics of office conversation and news sources and as the hearth around which household furniture is arranged. Children are raised with television, learning from it, entertaining themselves with it, and using it for obtaining beliefs about all aspects of social life.

This brings us to Winner's second anti-determinist argument. Most impacts of technology can be recognized as versions of earlier patterns. For instance, parents have always searched for ways to keep their children occupied for a few hours. New technologies do not bring completely new ways of being: "the very old patterns of human life are far too strong". But Winner explains that technologies do often transform the ways we fulfill these patterns, and therefore they can be called political. In other words, he argues for continuity and discontinuity at the same time.

Winner outlines two kinds of artifacts that "embody specific forms of power and authority" (Winner 1980). The first are instances in which the invention, design or arrangement of a technology becomes a way of settling an issue in the affairs of a particular community. In these instances, the function of the technology is to fulfill an active social role, as he explains with his famous example of the Long Island Bridges. Some bridges over the parkways on Long Island (New York) were built extremely low, some just 9 ft above street level. According to Winner, the typical design of these bridges had political consequences: it stopped black people going to the beach because (in general) black people were poor and therefore had to travel by public busses that were too high to pass under the bridges. The historical credibility of this case is questioned. Nonetheless, the example clearly illustrates the potential social role of technologies (Woolgar and Cooper 1999).

The bridges are designed to fulfill a social role and so here Winner seems to argue for voluntarism, but he also describes systems that require particular kinds of social structures for their functioning. Assembly lines for instance demand a strict coordination of labor. The pace of the work is determined by the rhythm of the machines, and so laborers must accept a rigid discipline that enables the machines to function (Winner 1980a). The tomato harvest machine—which cuts the plants, harvests the fruits, and stores the tomatoes in large trays—has changed the work and expertise needed for producing tomatoes. In the first decade of its existence, the harvest machine eliminated 32,000 jobs in the tomato industry. Further, this system calls for tomatoes that can endure considerable physical forces, and so the varieties became harder and sturdier. It happens that these tomatoes are also less tasty.

Winner's examples explain the importance of social arrangements that precede the use of technologies. These social roles are not necessarily the result of conscious conspiracies or malicious intentions in the designing process. Often unconsciously, unintended social roles emerge: most doors, stairs and curbs are real obstacles for elderly or less mobile people, hindering them from participating fully in social life. The things in themselves can assert "power and authority in human associations as well as the activities that take place within those arrangements" (Winner 1980a, p. 290). This is neither a defense of autonomism nor of voluntarism. He does not neglect the social influence on technological change, but recognizes at the same time that technologies can require specific social arrangements.

Bruno Latour has developed an approach to study how technologies and society influence each other. His work was one of the major drives behind the empirical turn in philosophy of technology and has co-shaped STS. His work combines the strengths of the four quadrants, as he does not fall into the trap of classing a priori one entity over the other. His question is not whether humans shape technologies or the other way around, since this leads to the mistake of a priory ranking one entity over the other. Instead, Latour offers an anthropological approach to study the interaction between humans and technologies on similar level (Callon and Latour 1992, p. 349).

> For us they [social constructivists] are exactly as reactionary as one who would start from an a priori unconstructed definition of nature in order to explain the settlement of controversies. On the contrary, we take as progressive any study that simultaneously shows the coproduction of society and nature.

His approach is called Actor Network Theory (ANT). This approach is anthropological in the sense that it adopts an observer's perspective to study actions like an ethnographer does. Initially, Latour disagreed with the label "ANT" (Latour 1999a), because he did not want to develop a theory (which is an inflexible thing in nature), but a methodology for sociology. However, in his later work, he claims ironically (Latour 2005a, p. 9):

> A name that is so awkward, so confusing, so meaningless that it deserves to be kept...I was ready to drop this label for more elaborate ones like 'sociology of translation', 'actant-rhyzome ontology', 'sociology of innovation', and so on, until someone pointed out to me that the acronym A.N.T. was perfectly fit a for blind myopic, workaholic, trail-sniffing, and collective traveller. An ant writing for other ants, this fits my project very well!

ANT basically comes down to describing the forces the human and technological actors exert on each other in the socio-technological networks (Latour 2005a).

With ANT, Latour offers a more radical approach to study the social role of technologies than Winner does. Latour argues, just like Winner, that some technologies are *deliberately* designed to encourage more "desirable" human actions. He provides the example of the blinking lights and the beeps that remind drivers to wear their seat belts. Other examples are easy to find. When the coffee is ready, some coffee machines make a loud noise, so one cannot forget to turn it off. A 10 cm high plastic pig is specifically designed to put in the refrigerator when one is dieting. It starts dancing and grunting wildly every time someone opens the fridge, as if to say: "eat some more and soon you will look like me!" In these examples it is clear that the technologies concerned were explicitly designed to influence our actions, but, as we will see, Latour argues that *all* technologies mediate human action.

Further, in Winners description, technologies and humans are still inhabitants of two different realms: the realm of the material and the realm of the social. By describing political artifacts, he worked to show that the material realm interfered with the social realm, and vice versa. Latour argues that there is no such thing as two realms: both humans and technologies are part of networks because they cannot be distinguished from each other based on their role in the techno-social realm (Latour 2002). In other words, individual humans, organizations, governments, companies, *and* technologies are actors, since they all exert a social force. Or, as Latour formulates (Latour 1992, p. 248): "it will become more and more difficult to trace the border between the empire of the human and the realm of technologies." Thus, both humans and technologies are called actors, actants or agents.

To discuss the first question—how can we best understand the social role of technology?—we can analyze Latour's approach by distinguishing three key features that characterize his work:

- Both humans and technologies are agents that fulfill a social role.
- The social role of technologies is best studied by looking at the human-technology associations that are formed, changed or broken off.
- The techno-social networks that make up our society are evolving overtime.

The next three subsections discuss these three aspects to obtain a deeper understanding of the answers ANT provides.

2.2.1 Human and Technological Agents

Latour discards what he identifies as one of the most important motives of modernism: the work of purification, in which nonhumans (objects) are described as inhabitants of the realm of nature and humans as part of the realm of culture (Latour 1991). Nature and culture are opposed to each other in many modernist philosophies, and so are objects and subjects. Latour claims that this dichotomy between object and subject does not hold (Latour 1992): humans are not atomistic, free and rational agents who can autonomously determine their own viewpoints and actions, and objects such as technologies are not simple instruments that refrain from exerting a social power. Instead, he describes both humans and technologies as agents (or actants) that are acting together in networks.

Latour expresses this by comparing a speed bump with a constantly present police officer. Sleeping policemen prevent speeding by physically forcing drivers to slow down (Latour 1994). Police officers become less involved when the regulation of speed is delegated to technologies: in other words, technologies fulfill the role of the police officers. Other examples are not hard to produce: a waiter can tell you where a non-smoking area is, and so can a sign. Your spouse can distract you from your work, and so can the television. Therefore, Latour calls both humans and technologies agents.

Here, Latour's point is not that technologies can replace people, but technological objects can "determine", "backdrop", "authorize, allow, afford, encourage, permit, suggest, influence, block, render possible, forbid, and so on" human action (Latour 2005a, p. 72). He shows that both humans and technologies have a similar social power, and therefore he believes that distinguishing the technical and the human dimensions in terms of means and ends, is a mistake.

Latour explains that both kinds of agents have their own "programs of action" instead of "ends", which consist of the agent's series of goals, steps and intentions. When human agents are not capable of realizing their programs of action, they fall back on technical artifacts. At that moment, the agent is taking a detour. The program of action of the technology mediates the human actions and the original ends change. Due to this detour, the initial agent will reach a new goal instead of the original. For example, a remote control is not a simple means to switch channels. It changes the ends as it leads to a completely different way of watching television. And as a consequence, it changes how people spend their leisure time: many hours are spent zapping through the channels, hoping for news items or for some entertainment.

In ANT, agency has four characteristics (Latour 2005a, p. 52): "agencies are part of an account; they are given a figure of some sort; they are opposed to other competing agencies; and, finally, they are accompanied by some explicit theory of action."

First, an entity is an agent if it acts, meaning that it has to make some difference to a state of affairs in the techno-social network. There should be some accounts, trails or transformations that can be observed. To find these transformations, two questions should asked: (1) Does it make a difference in the course of some other agent's action? And (2) Is there some trail that can be traced that allows someone to detect this difference? The "what if …" questions used in historical research are useful here: what would have been different if the technology had not been available? The answer to such a question can show that the technology has a social power that can be easily compared to the social power exerted by human agents.

Second, an agent should have some kind of figuration (Latour 2005a): a form or shape. This figuration can be abstract, like a character in a book, a corporate body or a central heating system, but if there is no figuration, there can be no agency. The Invisible Hand (Smith 1776) or other mysterious forms of social forces are not accepted in ANT.

Third, agents are "opposed to other competing agencies" which means that in pursuing their program of action, they will come across other agents with affirming, conflicting or transforming programs of action. Consequently, human agents cannot autonomously pursue their own ends. Latour describes that human actors are able to criticize other agents with normative comments. Artifacts obviously do not criticize other agents in the sense that they give verbal comments, but they are normative nonetheless.

Fourth, agents in sociological descriptions are often submitted to an explicit theory of action. Human agents are "able to propose their own theories of actions to explain how agencies' effects are carried over" p. 57. Technological agents do not have this capacity (of course) but Latour's point is that the metaphysics of the observer determines how the acts of both humans and technologies are perceived. This is equally true for human and technological agents, according to ANT.

So, networks consist of many agents or actors: each point in a network represents an acting agent, just like actors in a theatre play, but there is no real author of the play. Nor is there a script that fully determines the play. The only scripts are the "programs of action" of the different agents (Latour 2005a, p. 46):

> To use the word actor, means that it's never clear who and what is acting when we act since an actor on stage is never alone in acting. Play-acting puts us immediately into a thick imbroglio where the question of who is carrying out the action has become unfathomable. …Does the audience reaction count? What about the lighting? What is the backstage crew doing? Is the playwright's message faithfully transported or hopelessly bungled? Is the character carried over? And if so, by what? What are the partners doing? Where is the prompter?

Though both humans and technologies are agents in ANT, both are not regarded as exerting agency individually. Instead, agency is the result of the formation of associations between different agents (Latour 2005a, b).

2.2.2 Human-Technology Associations

Neither humans nor technologies shape the social, yet human-technology associations are responsible for change in the networks. Or, as Grint and Woolgar reformulate (Grint and Woolgar 1992, p. 374): "to break the network into its component parts—for the sake of analysis—is to misunderstand the nature of the event as a holistic process". Therefore, in ANT, agency is understood as the result of many human-technology interactions, and not as the result of autonomous human intentions.

Latour has developed two arguments to support his view. The first argument is "history as a whole" (Latour 2002): during the prehistory and history of humans, the emergence of humanity was interwoven with technical development. Every step we have made in our long history to develop different kinds of societies and cultures, relationships, skills and even our morality, was reciprocally connected to technological development. Technologies gave shape to human development and vice versa.

Second, Latour provides several extensive case studies. In *Aramis or the love of technology (Latour* 1996*)*, he explains how humans and technologies interact by describing a French R&D project on driverless metro cars. He wrote this book as a detective story, asking himself, who caused the death of Aramis (which is the name of the technology). To find the perpetrator, Latour treats humans as well as nonhumans as suspects and describes the interaction of the involved humans and technologies. This way, he shows that in the end the associations were to blame and not one human or nonhuman agent in particular.

To understand the interactions between humans and technologies, Latour describes techno-social networks of power exerting entities: humans and technologies exert power on each other in their aim to realize their own program of action. This "power-game" is obviously not limited to the interaction between one human and one technology. Agency is the accumulative result of four forms of mediation (Latour 1994): (1) translation, (2) composition, (3) black-boxing, and (4) delegation. Translation is about the detour a human agent takes when getting involved with a nonhuman agent (for example a technology). An agent, human or nonhuman, has a program of action that consists of series of goals, steps and intentions. If, for instance, humans are not strong enough to reach their goals, they often fall back on technologies. When this happens, the initial ends of the original agent are changed by the involvement of the new agent and both will reach their communal new goal.

With "mediation by composition" Latour explains that many actions and agents are already the result of a collection of humans and nonhumans. For instance, for manufacturing a car, one needs many technologies such as wheels, a carriage, an engine, electricity, and sheet metalwork. During the development of these technologies (which are all agents in ANT) many negotiations have taken place between different humans and nonhumans.

The third kind of mediation Latour describes is black-boxing, a process that makes the joint production of agents opaque. The acts of a driver (normally a human) are the result of the car, the driving lessons, the construction works that resulted in the roads, the traffic rules, the behavior of other driver-car compositions, and so on, and so forth. Simply "being in a hurry" is hardly an explanation for speeding: the other users of the road can make speeding possible or impossible and so can the road and the car. It is not even a necessary condition, since what counts as "speeding" is defined by the traffic rules. Some roads are designed for 100 km an hour, but because of the air pollution and noise, only 50 km an hour is allowed. The affordance of the road and the contemporary cars cause many people to exceed the speed limit.

The last form of mediation Latour distinguishes is delegation: they can take over human tasks. The speed bump that replaced the police officer is an example of this last form of mediation. This last form of mediation implies the first three forms.

Understanding these accumulating forms of technological mediation in a Latourian manner, leads to a huge shift in how we should perceive the world and "our" actions (Latour 1994, p. 40):

> … The relative shapes of actants and their ontological status may be completely reshuffled—techniques act as shape-changers, making a cop out of a bump in the road, lending a policeman the permanence and obstinacy of stone. The relative ordering of presence and absence is redistributed - we hourly encounter hundreds, even thousands of absent makers who are

remote in time and space yet simultaneously active and present. And through such detours, finally, the political order is subverted, since I rely on many delegated actions that themselves make me do things on behalf of others who are no longer here and that I have not elected and the course of whose existence I cannot even retrace.

2.2.3 Changing Associations

From this quote, we may conclude that the associations between humans and technologies are not fixed in time. Instead, the network is constantly changing, and we can only understand the social role of technologies in the evolving network. Agency is a-local, meaning it cannot be pinpointed to a specific site (Latour 2005a). Agents are made to act by many other agents, and so, they are not the source of what might seem to be their own actions. This implies that the origin of action is uncertain in the actor-network. As Latour continues (Latour 2005a, p. 46):

> By definition, action is dislocated. Action is borrowed, distributed, suggested, influenced, dominated, betrayed, translated. If an actor is said to be an actor-network, it is first of all to underline that it represents the major source of uncertainty about the origin of action.

Thus, the techno-social networks should not be understood as a collection of stable, fixed relationships. On the contrary, networks are constantly changing as the multiple agents exert power on each other. This explanation of what it means to be an agent in ANT presumes a metaphysical view that can be compared with pinball: the ball's program of action consists of the inclination to roll down the slope. The player's program of action is to collect points and win the game by keeping the ball from disappearing off the bottom of the playfield. The flippers are used to redirect the ball upwards, but a skilled player may also use these flippers to hit scoring targets. Note, the flippers are placed in such a manner that the ball can always slip between them if rolling down from a straight angle. Beside the ball, the flippers and the player, many more "agents" fulfill a role in unfolding the game: the plunger provides the metal ball with speed when it is launched into the playfield. Electrical switches detect the speed and direction of the ball, which enables the scoring mechanism to count points and to grant extra balls. Additionally, there are active and passive bumpers, kickers, holes, slingshots, spinners, ramps, stationary targets, magnets, rollovers, bullseyes, drop targets, saucers, switches, gates and so on, and so on. All these devices constantly redirect the ball and give points. The software makes the various elements respond differently in distinct stages of the game. Extra points may be added or extra balls launched (one after the other, or simultaneously) when excellent balls are played.

Just as the ball, an agent in ANT is constantly redirected and "made to act" by many other actors that "swarm toward it" (Latour 2005a, p. 46). Furthermore, just like the playing field, the techno-social networks are changing in time. Describing the situation of a pinball game at one point in time is of little use if one wants to understand the game. Nor is it enough to know the few rules of the game: for playing the game successfully, one needs to understand how the multiple elements of the pinball machine interact. The same holds for the networks: if all actors exert power on each other—constantly changing the networks—it is of little use to only provide a description of the situation at a certain point in time, nor is it enough to know the rules that are embedded in that network. For understanding and influencing the network it is important to unfold how associations are formed, altered or broken off, and how these associations lead to action.

Summarizing, we can say that ANT provides a clearer view of the social role of technologies. ANT explains: (1) that both humans and technologies are comparable in exerting social forces, and that therefore they should both be studied as agents; (2) that the social force of human and technological agents is best studied by looking at the human-technology associations in which the agents express their programs of action; and, (3) that because the networks are fluid, we can only understand techno-social change by studying the evolving associations.

So, Latour offers a means to study how technologies alter our actions without falling into technological determinism or autonomism: it teaches how we are mediated in our action by technologies when striving to realize our programs of actions. Though exerting agency as defined by Latour is compared with playing pinball, other actors are not fully in control either. As soon as the player stops, the ball will quickly disappear between the flippers. But the player has several means—the flippers and nudging (tilting the table)—to prevent the ball from going down.

The aim of getting a clearer view on the social role of technologies was to deal with the question of whether we can take responsibility in the sense that we can enquire what the possible social role of a technology will be and adjust our actions according to the evaluation of this enquiry. Does ANT help to take such a responsibility?

2.3 ANT and Responsibility

At first glance, it seems that ANT contributes substantially in accepting responsibility for the social role of technologies. It provides an approach in which the social role of technologies can be studied without accepting a priori dominance of one entity over the other. ANT makes the social power of technologies intelligible by comparing them to the social force of humans. Latour explains that we can understand the actions of technologies by studying the agency as erupting from the interactions between human and nonhuman agents. Agency can be recognized as the result of different agents who are acting simultaneously. The associations between technologies and humans are comparable to the relations between humans, and agents are made to act in a given way by the associations with many other agents—human or nonhuman. Therefore, the human agent should not be understood as the source of action, nor does the nonhuman agent fulfill this role.

With this, Latour provides an approach to study the role of technologies, incorporating the useful insights related to all four quadrants of the matrix. ANT explains why we are not as autonomous as instrumentalism and voluntarism claim, while it explains that we are not fully determined by technological developments. ANT provides a complex, but useful approach that helps to describe techno-social change. Having means to describe these changes is crucial when discussing responsibility issues concerning the social role of technologies. So, on this level, ANT seems to say "Yes" to the responsibility question since it enables us to alter our program of action and affect the developments.

However, when reconsidering the metaphysics behind ANT, the conclusion that there are so many interacting forces that the origin of the action is dislocal, frustrates the possibility of taking responsibility. How can human agents take responsibility if they are like the ball on the pinball table, wanting to escape through the hole in the middle, but constantly being smashed around by all kinds of other interacting elements?

In ANT, humans are not the only entities with a program of action. Even in the unlikely event that all humans would share the same program of action, we should be aware that technologies are also directive in the socio-techno networks which will alter the outcomes of actions. Humans and technologies are all described by ANT as dots in a network exerting power on each other, and so the networks represent a substantial collection of interacting forces presenting action as dislocated.

This view results from Latour's wish to use non-discriminating language for his description of how humans and technologies interact in a manner that is as detailed and neutral as possible. He wants to treat humans and non-humans as equal as possible regarding their social roles. But is that fair? Collins and Yearley were the first to point out that Latour assigns too many features of subjects to objects, and vice versa (Collins and Yearley 1992). According to Collins and Yearley, Latour is giving agency *back to nature* because his language introduces symmetry between humans and nonhumans; he anthropomorphizes technologies and, at the same time, objectifies humans. Latour would not deny this (see the following chapter), but he does not accept that this is problematic.

But the symmetry thesis has rather problematic consequences for responsibility questions. Aaron Smith introduces the problem of locating moral responsibility in ANT as follows (Smith 2003, p. 182):

> Two figures stand twenty paces apart staring silently at one another. The first one is an American phenomenologist. The other is a refugee from the science war. Both are armed with knowledge of philosophy and a few other mundane technical artifacts and actants. The tension breaks as Ihde reaches for his gun. … Latour slumps to the ground and dies a seemingly quick and painless death. … The townspeople look at each other with great confusion.
>
> 'Ihde just killed a man.'
>
> 'But Ihde did not kill him. The bullet caused the wound that ended Latour's life. Let's take the bullet to jail.'
>
> 'No, wait. The gun should go to jail because the gun is what gave the bullet the force to make the wound that ended Latour's life.'

Smith asks: if agency is the result of human-nonhuman associations, and therefore a-local, then how to deal with responsibility issues? If the action is dislocal, so is the responsibility, and this creates an important problem for responsibility issues.

Smith's joke has its roots in the US public debate on guns. The well-known National Rifle Association (NRA) bumper-sticker, "Guns Don't Kill People, People Kill People" is a reaction toward the slogan "Guns Kill People" used by people who plead for stricter legislation on weapons like guns. By restricting the availability of guns, this movement aims to prevent accidents, suicides, manslaughter, and murder. As Latour explains, they argue rightly against the instrumentalist slogan of the NRA. Having a gun is a necessary condition for becoming a gunman (Latour 1999b); a gun cannot shoot by itself, nor can a human shoot without a technological artifact like a gun. The *association* between human and gun or rifle makes a shooting possible. This seems logical, and the NRA would not disagree, but ANT rightly points out that there is more to it: the availability of a gun changes people's actions on a deeper level. For instance, when people have a gun, they feel more confident to try to catch a burglar in the middle of the night. With a gun in one's hand, one can feel more daring, more aggressive, more in control of the situation. Therefore, it is wrong to say that Ihde is the only origin of action (the gun, the bullet, Latour, and the inciting crowd also play a role).

But if we believe that agency is spread over many power-exerting agents in the network, how then can we take responsibility for the social role of technologies? The following chapter discusses this problem in more detail, and set the agenda for the rest of this study.

Chapter 3
Actor-Networks and Taking Responsibility

The first question of this study stressed that for dealing with the problem of whether we can take responsibility, it needs to become clear what it actually means to say that technologies play a social role. As clarified in Chap. 2, Actor Network Theory (ANT) teaches three important lessons: first, in exerting social power, the actions of humans and technologies are comparable. Second, the origin of action is dislocal, meaning that not just one human or technology determines the outcome, but that agency is the result of interactions between multiple humans and technologies. Third, the techno-social networks are constantly evolving as the human-technology associations are continuously formed, changed and broken off.

However, these three lessons pose serious challenges for engineers and other practitioners who want to take responsibility for the mediating role of technologies: how can one take responsibility if technologies have some power too in the sense that the interaction between humans and technologies determine the continuously changing outcomes? This chapter explores the difficulties that arise for engineers if ANT is believed to be correct.

3.1 ANT and Responsibility

To what extent does ANT empower people to take responsibility? Or to repeat the second question: can practitioners still take responsibility for the social role of technologies? As argued in Chap. 1, a particular notion of responsibility is at issue here. In relation to responsibility, most philosophers focus either on liability or duty. In the former, "being responsible" means being rightfully the target of responsive attitudes: you have done some right or wrong for which you ought to be praised or blamed. In the latter, it is stressed that you are being responsible for something in the sense that it is your duty to do certain things. But here the issue is different. It concerns responsibility for the social role of future technologies. Praise and blame are terms that are

K. Waelbers, *Doing Good with Technologies: Taking Responsibility for the Social Role of Emerging Technologies*, Philosophy of Engineering and Technology 4,
DOI 10.1007/978-94-007-1640-7_3, © Springer Science+Business Media B.V. 2011

commonly used retrospectively, after the actions have taken place and when consequences have become clear. Duties can be assigned prospectively, but the problem is that assigning duties is only feasible when it is clear what must be done. When discussing actions for new technologies, the best thing to do is often unclear. So instead of liability or moral obligation, the issue here is whether someone is able and willing to (a) enquire what the possible social role of a future technology, (b) evaluate this possible social role, and (c) adjust one's actions according to this evaluation.

We can analyze these three points with help from the conditions for taking responsibility that are often addressed in ethics. This enables us to see how Latourian lessons relate to responsibility questions, and to see which lessons of ANT should be preserved.

However, an ethical analyzes of ANT is controversial, since the starting points of most ethical theories are rejected by Bruno Latour. Almost all schools of ethics argue that people are moral subjects whose actions should be distinguished from those of technologies. Kantianism (deontology) argues that people can do the right thing using rational reflection on moral principles. Utilitarianism teaches how to decide what a right action is by finding out what brings the most happiness to the most people. Virtue ethics explains that we can learn to be good from living with our parents, teachers, family, colleagues, friends and governments. Regardless of the substantial differences in approach between these schools, they have two essentials in common that are rejected forcefully by Latour. First, they presume that people will act good, right or virtuously once they understand the appropriate moral arguments. In other words, many ethical theories presume that people will act in a desired manner when they understand the arguments of why it is desirable to do so. Second, ethical theories do not commonly consider technological artifacts—or any other object for that matter—to be a substantial part of the moral sphere, in the sense that they are not considered to play a social role (Smith 2003, p. 183):

> Most ethical theories leave the status of technology and technical artifacts as neutral or playing a small, unimportant role. A deontologist, consequentialist or virtue theorist would argue that it just does not make sense to bring the objects into the discussion.

ANT rejects these two essentials of ethics and argues that people are not atomist, rational subjects who can freely determine their own course of actions, because technologies are also part of the moral domain. ANT relocates morality (Latour 1992; Swierstra 2000; Verbeek 2005a) which means that morality can be found in humans *and* things. Artifacts are moral because, just like people, technologies can direct human actions in certain directions. For example, a well-designed stool can make you adopt a bodily posture that physiotherapists recommend. An electronic agenda helps one to be mindful of the birthdays of your friends and family. Such technologies are even better in moralizing than humans, Latour argues, for people do not alter their actions because they are convinced of rational moral arguments (Latour 1992, p. 232):

> In spite of the constant weeping of moralists, no human is as relentlessly moral as a machine. ...We have been able to delegate to nonhumans not only force as we have known it for

> centuries but also values, duties and ethics. It is because of this morality that we, humans, behave so ethically, no matter how weak and wicked we feel we are.

Latour observes that ethicists wrongfully expect people to do the right thing because of the validity of their moral arguments. His observation is not hard to confirm: even if it is for their own good—like adopting a healthy lifestyle—people often do not adjust their actions simply because they understand the arguments to do so. To substitute for the lack of efficacy of human morality (Latour 1992, 2005a), Latour adopts a more Machiavellian approach. He argues that we need "other kinds of allies"—in other words, we need technologies to help us—if we want to make sure that people act in certain manners (Latour 1987a, 1988). He explains (Latour 1988):

> Machiavelli was looking for a point of view from which all the contradictory advices given to the Prince will make sense: stay in power a bit longer in spite of the vagaries of fortune. The point I would choose is rather this one: make your environment such that whatever other human or non-human actors think or do, they are either kept at bay or else they help strengthen your position, making the world safer, more predictable and more enjoyable for you.

Latour—being a democrat—does not plead for a society in which we are all dictated to technologies, but he aims for a political system in which the social power of technological artifacts is acknowledged (Latour 2005). With this move, he explicitly rejects the two most dominant ethical theories deontology and utalitarianism (Latour 1995b) because he does not agree with the two points of departure. For Latour, humans are not autonomous and technologies are politically relevant.

Using criteria of moral philosophy to analyze the mainly sociological ANT is controversial, but the aim here is not to criticize ANT with traditional ethical theory. The aim is to develop a moral view for taking responsibility for technologies that incorporates important elements of ANT. Latour is arguing against those two presuppositions of the main ethical theories, but he is not arguing against the desirability of taking responsibility. The conditions for responsibility presented above are not just deduced from moral theories: they are also part of our common moral beliefs about what one needs for taking responsibility. Furthermore, the conditions are also recognizable in engineering practice (Swierstra and Jelsma 2005, 2006). For being able to take a certain responsibility, four criteria are essential:

- The actor is imputable
- The actor acts according to his or her free will (the actor has the capacities of reasoning, feeling emotions and acting intentionally)
- The actor is at the origin of the action (he or she can make a difference)
- The actor can foresee the consequences

Of course, the first criterion is less relevant when discussing ANT, but when analyzing the key notions of ANT with the help of the remaining criteria, three worries arise (see also Waelbers 2009b). First, can people act freely and willingly when technologies alter people's programs of action? Second, can people be the cause of their own actions if agency is the result of human-technology associations? Third, can people act knowingly in a black-boxed and evolving techno-social network? The next three sections explore these questions.

3.2 Acting Willingly

What happens if we compare the first condition for taking responsibility with the first lesson of ANT? The second condition for taking responsibility is that one should be able to act willingly. This point has been made by Immanuel Kant, when he explains that we can only act good if we act according to our good will (Kant 1785/1993). Most moral theories explain that only people are agents, since they can act according to their own will. The first lesson we can learn from ANT is that humans and technologies play similar social roles but, obviously, technologies do not act willingly. Therefore, Latour needs a concept that can be used for describing the actions of both entities, and he chooses "programs of action" for this aim. This concept enables him to compare the actions of technologies with the actions of humans. It offers a redefinition of agency that enables sociologists to study the human-technology relationship without prior prejudice about dominance of one entity over the other. Such an approach helps to overcome the traditional opposite between sociology and philosophy debates on the social role of technologies, and to move toward the study of the social role of particular technologies (the empirical turn).

In Latour's anthropological framework, acting willingly is not important for understanding the social because the outcomes of the interactions between the different social groups are objects of study. Latour, being an anthropologist, wants to display the importance of the social power of technologies by explaining that humans and technologies interact and that the social is the result of these interactions. He understands agency in a specific anthropological manner: "an actor is what is made to act by many others" (Latour 2005a, p. 46).

The concept of agency in ANT thus has a different focus than is common in moral philosophy or, for that matter, in daily practice. In ANT, the outcomes of action (or interaction) are important, while moral philosophy focuses on the intentions behind actions. In moral philosophy, human action is explored not so much with reference to action outputs but with action inputs. Three human capacities that provide reasons for actions—the capacity to reason, to form desires, and to have intentions—are essential for responsible agency (Mele 2006) and only with one or more of these capacities, is acting willingly possible. Consider briefly introduce the meaning of these three capacities in relation to ANT.

The Capacity of Reasoning Moral philosophy argues that agents are persons who can employ their capacities to reason in order to act willingly. People are moral agents because they have the ability to reflect on their options, to make assessments and to adjust actions accordingly. To be able to act willingly (and therefore to be able to take responsibility) the capacity to reflect about one's actions is essential. Or to put it simply, to act responsibly in an ethical sense, one has to be able to make conscious decisions about what to do.

As long as we are not discussing futuristic forms of artificial intelligence, technologies do not have this capacity and therefore the concept of agency is incompatible with artifacts (Noorman 2008). Latour acknowledges this difference between humans and nonhumans when he argues that human agents, unlike nonhuman agents, are able to criticize other agents and that they formulate theories on their own agency (Latour 2005a). But in ANT, these characteristics are not necessary conditions for wielding agency. Moreover, ANT argues that this capacity does not make much difference for moral actions. As mentioned, Latour argues that people do not alter their actions because they are presented with good arguments (Latour 1987a, 1992, 2005a, b). He argues that the moral capacities of technologies make us behave one way rather than another, and not so much our moral beliefs. This leads him to conclude that despite the dissimilarities between humans and non-humans regarding their ability of reasoning, the outcomes are similar.

Latour is right in stating that many actions are unreflective and that it is also often the case that people do not adjust their actions according to their deliberations even though they should. Yet, from an ethical point of view, this response is only partially satisfying. The argument that people often behave in undesirable manners does not mean that they should not act otherwise. Additionally, many people also act from the desire to do good: if not, how for instance would we explain altruistic actions such as the huge amounts of money people donate to charitable institutions? A description of how the actors behave in the networks should also take into account the capacity of reasoning.

The Capacity to Have Desires and Feel Emotions Second, to act willingly, human agents need the capacity to have desires and to feel emotions. When discussing the moral status of animals, Jeremy Bentham (1748–1832) pointed out that being able to suffer is first and foremost important for being morally relevant (Bentham 1789/1996, p. 283): "the question is not, Can they *reason*?, nor Can they *talk*? but, Can they *suffer?*" Utilitarianism has explained that if you are able to feel pain or to enjoy comfort, others should pay respect to your wellbeing. Artifacts do not have this capacity, so therefore they are not morally relevant entities according to utilitarianism (Swierstra 1999).

More recently, Peter-Paul Verbeek pointed out that this argument is too limited. When talking about the human-technology interaction, it is important to distinguish between moral agents and moral patients (Verbeek 2005b). In other words, the question of whether something is able to feel pain or happiness does not say much about the question of whether something can be moralizing. Consequently, Verbeek explains that technologies are those entities that are moral agents, but not moral patients.

Though it is unmistakably true that technologies can be moralizing, they are not moral agents in the same sense as humans. Humans have desires and emotions which provide them with the capacity to empathize, that is they can feel sorry or happy for someone else, and this capacity can be essential for exerting moral agency (and thus for taking responsibility). Technological artifacts such as razors, light bulbs and espresso machines lack the capacity to have desires, emotions and empathy all together. So, technologies may exert a moralizing role, but they are not moral entities in the same sense as humans.

According to some philosophers, having desires (Davidson 1963) and feeling emotions (Roeser 2005) is morally relevant by itself. Many philosophers argue the opposite when they say that the capacity to reason should be used to affirm or correct desires and emotions. Regardless of the viewpoint that is chosen, the ability to desire something and to feel emotions is acknowledged by almost all psychological theories as being essential for being a moral agent who can act willingly.

But desires and emotions are not a subject of enquiry for ANT (or they are described as social forces), even though they are an important distinction between how and why human and technologies act. Describing humans or technologies in terms of their "program of action" can be done without referring to anything such as desires, emotions or empathy. For the sake of symmetry, ANT also describes human agency in this "cold" manner, so it remains possible to compare the social force of humans with the social force of technologies.

The Capacity to Have Intentions Third, acting willingly means that someone acts intentionally; people are purposive beings, who set goals for their actions. Ethical theories argue that for understanding, evaluating and responding to human action, it is important to understand the reasons or (moral) intentions behind the actions. Ethics is about describing *what* to do and *why* to do it: questioning *why* something should be done is at least as important for taking responsibility as knowing *what* should be done. The notion of intentionality addresses this "why" question. Humans intend to do certain things and to intentionally refrain from other actions. In other words, they do things on purpose. Even when we give up the idea that full autonomy is needed for moral agency, we cannot give up the idea that moral agents needs some form of awareness which enables them to act purposefully. Consequently, in acting willingly, humans cannot be compared with nonhumans. When analyzing socio-technological developments, we need to incorporate this distinction (Collins and Kusch 1998; Collins and Yearley 1992; Pickering 1995; Smith 2003; Swierstra 1995).

Latour treats humans as "teleological beings" but he does so in a manner in which technologies are teleological as well (Khong 2003): both have their own programs of action. A description of a program of action is for instance "John aims to give fifty euros to Greenpeace". Such a sentence resembles the line "the eco- program on the washing machine aims to make us wash cheaper and more environmentally friendly". Such a description of a program of action focuses on the course of actions and does not address a person's moral intentions. John could for instance be willing to donate the money because he lost a bet with his wife, or because he wants to compensate for flying so often, or because he believes it is everyone's duty to preserve nature. ANT does not question the intentions behind the program of action: it just studies what people and technologies aim for.

Latour has chosen an output-oriented approach because he aims to describe techno-social networks as neutrally as possible. He does not want to do "interpretative sociology" because then the "scene of an individual subjective actor having 'some leeway' 'inside' a larger system will be reactivated" (Latour 2005a, p. 206). In other words, Latour fears that as soon as intentionality is presupposed, the atomist autonomous subject is reintroduced via the back door, and the social will be explained by the wrongly understood rational intentions of humans (Latour 2005a, p. 61):

> But an 'interpretative' sociology [meaning sociology which aims to interpret human intentions] is just as much a sociology of the social than any of the 'objectivist' or 'positivist' versions it wishes to replace. It believes that certain types of agencies—persons, intention, feeling, work, face-to-face interaction—will automatically bring life, richness and 'humanity'. This belief in the 'lived world' is a nice case of 'misplaced concreteness' to use Whitehead's term.

3.2.1 Debating Intentions and Responsibility

In understanding what it means to be a moral agent who can act willingly, Latour has a different approach than ethics, since he has a different aim than ethicists. Therefore, he adopts an unusual concept of agency. It is not uncommon that people in different fields of study use dissimilar concepts to address different issues. Differences in research aims between sociologists and moral philosophers lead to a differences in their use of concepts (Hoeyer 2006). Every discipline establishes its own aims for research and to adopt available concepts in their own particular ends. Nonetheless, this discrepancy is problematic. Focusing on reasons is not enough for understanding why people act the way they do: social forces that influence the initial action are too important to ignore. But, social forces do not fully determine the intentions, and the intentions behind human agency are important for explaining the techno-social networks and for taking responsibility in those techno-social networks.

Latour of course recognizes the differences between intentional and unintentional actions (Latour 1992, p. 353):

> One way to do this [research] is to extend our principle of symmetry to vocabulary and to decide that whatever term is used for humans, we will use it for nonhumans as well. It does not mean that we wish to extend intentionality to things, or mechanism to humans, but only that with any one attribute we should be able to depict the other by doing this crisscrossing of the divide, we hope to overcome the difficulty of siding with one, and only one, of the camps.

And (Callon and Latour 1992, p. 359):

> the intention was not to say that scallops have voting power and will exercise it, or that door closers are entitled to social benefits and burial rites, but that a common vocabulary and a common ontology should be created by crisscrossing the divide by borrowing terms from one end to depict the other.

So, Latour's argument is not that people lack intentions, but he is arguing that when studying human and technological agency, it is better not to theorize about intentions, since this would lead to false ideas about the social. Analyzing actions in terms of moral reasons obviously does not contribute to this aim: it will automatically remind us of the human input and will present the technologies as socially dead objects. So "we are not talking of intentions here, but of use and of effects" (Callon and Latour 1992, p. 358). The language describing agents (human and technological) as entities that exert social force to realize their programs of action, is far better suited for showing that the social roles of technologies and humans are comparable.

He explains this statement by comparing human interactions with billiard balls that hit one another. The agency people exert in a pub is much like playing billiards: people who are looking for a "familiar face in the smoke-filled room" just hit upon someone, or not. But the question is whether Latour is correct here. Of course, socializing in a bar is often not planned, but that is not the point. For understanding social interactions (and especially for being able to adjust one's actions), it is also important to make the intentions behind the human actions explicit. When an irritating man approaches a woman in a bar, she often bounces quickly to someone else in the room. But, why does the woman think he is irritating? Is she put off by his too obtrusive behavior, his bad smell, or his overdeveloped interest in soccer?

Latour has chosen to limit his studies to the level of forces, because he aims to be as neutral and realistic as possible in describing the techno-social networks (even though he does not believe in neutral science). But one can wonder whether his understanding of the social is really neutral since he refuses to look at the moral motives. As Tsjalling Swierstra has pointed out (Swierstra 1992), the vocabulary of Latour is rather war-like. He quotes a long list of verbs from *Science in Action* (Latour 1987b): "competitor" (p. 5), "game" (p. 6), "awed" (p. 6), "strategic advice" (p. 9), "struggling to recruit" (p. 13), "arms race" (p. 26), "win" (p. 29), "force" (p. 30), "allies" (p. 31), "attack" (p. 36). In other works of Latour—works that focus more on the social role of artifacts—this language is equally prominent.

Talking and thinking about the world in such a manner resembles a Machiavellian world-view, as becomes also readily clear from "How to Write 'The Prince' for Machines", an older article of Latour (Latour 1987a, 1988). But this is not a truly neutral and realistic view since it excludes explanations that are also essential for understanding the social. Latour notices, for instance, that Christianity imposes moral rules on other people, while in ANT these are understood as external forces. ANT ignores that internal moral considerations influence someone's identity and actions.

In a social setting, people expect social or moral explanations that tell them what to do. Consider the following case. To get attention for an after-dinner speech, someone gently taps with a knife on her wineglass. The audience politely turns their chairs in her direction and look expectantly. Who is making that sound, an old, short-sighted man on the far end of the table asks. According to ANT, the correct answer would be that the interaction between the three actors produced the noise together: the woman, the knife, and the wineglass are mutually responsible. The argument is simple: all three elements are necessary since without the glass, the knife or the woman, it would not have been possible to attract the attention of the audience in this manner. But the man in the audience would be rather puzzled if she, standing alone at the other end of the table, replied to him "the three of us do, sir." Obviously he wants an answer that addresses the purposive agent, which would be her and not the wineglass or the knife.

She produced the sound for a reason, since she wanted to deliver a speech. The knife and the wineglass were used for realizing her wish to attract attention. Of course, the volume and beauty of the sound depends on which kind of glass and knife she uses. And, it sometimes also occurs accidently, someone tapping a wineglass with a knife. In such a situation, the attention of other people may also be attracted. But the mere fact that an accident can lead to the same situation as an intentional action does not imply that we should treat them the same: when the old short-sighted man gets a proper answer to his question, he understands the social event and he will be more likely to be quiet and listen to the speech.

Understanding and evaluating human actions is philosophically interesting: we need to be able to respond properly to other people's actions. If you want to understand and evaluate human actions—varying from buying flowers and sending money to Africa to committing a crime of passion or stealing a car—it is important to explicate the personal or moral intentions.

Additionally, being a responsible moral agent also includes caring for the welfare of other people, animals, and the environment. Being a moral agent is what makes us human. Beliefs on how to be and do good give meaning to our actions and to our lives. This does not imply that everyone is always directed toward doing good: most of the time people just behave according to their customs. Selfishness, ignorance, and cruelty are common. But often people do reflect how they want to live and how to do good in particular situations. These reflections are ethical in the sense that they include not only prudent arguments, but also the interests of others. Fundamental indifference to the welfare of others and exclusive focus on one's needs is in psychiatry seen as a clear sign of mental illness (the DSM IV personality disorders) because it obstructs "normal" interaction with people.

People have beliefs about what it means to have and live a good life that are at the base of our cultures and reflect the quality of life in our societies. As mentioned, this aspect of human agency is simply too important to ignore in an anthropological philosophy that aims to understand the social (Latour 2005a). By ignoring the substantial differences between humans and technologies, the social sphere becomes disappointingly shallow and meaningless. Ignoring internal motives is equally problematic for taking moral responsibility.

In taking responsibility it is important to reflect on what is morally desirable, and so, not limit the focus on power games. Taking a forward-looking responsibility means to ask the question: "how can I promote a flourishing life for myself, other people, other animals, and nature?" This implies that without including aims, desires, and intentions in a social analysis, it does not make much sense to talk about moral responsibility. If human agency is described in the same terms as the social role of technologies—meaning only by referring to their "program of action" and without any reference to aims, desires, and intentions—their capabilities of taking moral responsibility are underestimated. To take a forward-looking responsibility, people need to reflect on their reasons or motives for action (see also the following chapter). If these starting points are ignored, and the focus remains on unreflective programs of action, a plea for taking responsibility is not likely to be successful.

3.3 Causation and Mediation

The third criterion explains that for taking responsibility, people should be able to act themselves: they should be the cause—or at least a major cause—of the consequences. Translated to everyday language, this condition says it would be unfair to ask people to take moral responsibility for something they did not do. Of course, not doing something—like not telling the whole truth—can also be an act for which responsibility should be taken. One can wonder whether this causal condition can be fulfilled in the engineering practice, since these activities involve many people: it is simply unclear whose actions caused what effect. In the philosophy of management this problem has been put on the agenda with the phrase the "problem of the many hands" (Thomson 1980).

The problem of causality becomes even more substantial when we take the second starting point of ANT seriously (Waelbers 2009b). It is not simply the case that even more "hands" are added to the problem of the many hands: the second starting point explains that agents alter each other's programs of action. ANT understands social change in terms of agents who exert social force on each other in an attempt to realize their programs of action. This is a field of study that focuses on the productive social power that is exerted in the techno-social networks. ANT shows that the very results agents aimed at are mediated by other agents. Consequently, ANT understands the origin of action as dislocal.

3.3.1 The Implications of Dislocalizing Action

Claims that the origin of action is dislocal and that agency is the emergent result of many human and nonhuman agents collaboration in action, raise the question of whether we can still be the cause of action (Waelbers 2009a). Latour's quickest reply is "No" we are not the origin of action (Latour 2005a, p. 46): "an 'actor' in the hyphenated expression actor-network is not the source of an action but the moving target of a vast array of entities swarming toward it." He does not want to distinguish between different kinds of causation since he aims for a neutral approach: he does not want to introduce a priori all kinds of frameworks in his studies for this would "corrupt" his observations.

His refusal to distinguish different kinds of causes has led Collins and Yearley to argue that he wrongfully anthropomorphizes technological artifacts and, at the same time, objectifies humans (Collins and Yearley 1992). Consequently, Latour's understanding of the social is deficient. Latour admits to anthropomorphizing technical objects in at least three different ways (Latour 1992, p. 235). First, he argues that technical objects can be like humans in the sense that they fulfill human tasks. Robots in assembly lines are a good example of this form of anthropomorphism. Second, technological artifacts are shaped by humans: they are anthropomorphic in the sense that their forms and functions (also) have a human origin. Third, Latour argues that artifacts mediate humans in the sense that they moralize them just as humans moralize one another. The door groom example is anthropomorphic in all three meanings (Latour 1992, p. 234): "first, it has been made by humans; second, it substitutes for the actions of people and is a delegate that permanently occupies the position of a human; and third, it shapes human action by prescribing back what sort of people should pass through the door."

But of course, these forms of anthropomorphizing are not very problematic and the critics refer to another, less innocent form of anthropomorphizing: they accuse Latour of ascribing typically human characteristics to technological artifacts (Khong 2003) and consequentially spreading moral responsibility too thinly (Noorman 2008).

In studying the social, Latour redefines agency in a purely causal sense: something is an agent if it redirects other agents. In other words, if an entity exerts social power, then that entity is an agent. Such a view leads to spreading responsibility among many human and technological agents (Waelbers 2009a). Latour is explicit on what his explanation of causation implies for responsibility issues (Latour 1999b, p. 180):

> "actor-actant" symmetry force(s) us to abandon the subject-object dichotomy, a distinction that presents the understanding of the actual operation of human-technological associations. It is neither people nor guns that kill. Responsibility for the action must be shared among the various actants.

Since agency is the result of many negotiations between various human and non-human agents, the question of "who caused the action" becomes hard to answer. We need to "trace back" the formation of the various human-technology associations, and include the role of each actor in the analysis. When conducting such an actor-network analysis, often the conclusion follows that many human and technological agents involved share the causal responsibility. But causal responsibility is not the same as moral responsibility (Lenk 1993; Ropohl 1993). Causal responsibility only explains which action caused which reaction: it explains mechanics. The lightning can be causally responsible for setting a farm on fire, but it is not morally responsible. Moral responsibility goes beyond this point: it asks questions about why someone does or did something, whether this purposive action and the outcomes are morally desirable and whether the person could and should have acted differently to alter the outcomes. Latour is not talking about moral responsibility at all and focuses solely on causality.

Artifacts may be causally responsible, but of course they are never held morally responsible (Swierstra 1999). We do not drag a car to court because it was part of a traffic accident. We never expect a sincere apology from technological artifacts, because they are just mechanical causes: they could not have done anything differently, nor can they feel sorry for what they did. There is no reasonable "why" question you can ask when the lightning sets fire to the roof of your house since the lightning does not strike purposefully and cannot decide to strike on your neighbors' home instead. By contrast, people can act deliberately, and so be held responsible.

Although moral responsibility implies causal responsibility, causation does not work as simply as common sociology and philosophy often assume. As Latour explains, the origin of action is dislocal. Latour introduced four accumulating forms or aspects of mediation to explain how this works (Latour 1994): (1) translation, (2) composition, (3) black-boxing, and (4) delegation. The first aspect is the translation of action that takes place when a human agent employs a singular technological artifact. In this process of translation the initial ends of the original agent are changed by involving the new agent and both will reach their communal new goal instead of the initial agent's original goal. Latour speaks of agents merging in action. This merging takes place many times in a row, artifacts such as televisions, cars, or buildings are made up of multiple other artifacts and are the results of many negotiations. This leads to the second aspect: mediation by composition. Latour explains that many agents are already the result of a collection of humans and nonhumans. To address the opaqueness of all the interrelated social forces that act in the Matryoshka-like networks, Latour uses the term black-boxing. The doings of technical artifacts are subject to black-boxing, a process that makes the joint production of actors and artifacts entirely opaque. One does not see the different components of the artifact, and only by malfunctioning, does one realize the existence of the black-box. For instance, when an automobile does not start, one needs a mechanic to open the black-box that is the car. Most black-boxes themselves consist of many other black-boxes. When the black-box "car" is opened by the service engineer, the engineer may conclude that the solenoid is broken, but repairing the solenoid herself is often not feasible. It simply needs replaced as a whole. Summarizing, mediation takes the shape of composition: a technological artifact or a business is composed of different humans and technologies. These compositions become less visible and present themselves as singular black-boxed agents.

Will practitioners accept a proactive responsibility for the Matryoshka-like effects that result from Latour's mediation theory? Consider the example of contemporary bio-industrial chicken farms. On the one hand, many technologies and many humans have played a role in their development. Electricity, antibiotics, vets, insurance companies, banks, the government, farmers, chickens, architectural technologies, consumers, and so on, and so on, all interacted over a period of since 40 years (Waelbers et al. 2004) resulting in the current large rows of chicken. These cages have, however, become one system—or one agent, as Latour would argue—whose existence has become the standard. On the other hand, interviews with scientists show that this problem is only partially recognizable in the engineering practice (Swierstra and Jelsma 2005, p. 208):

> Interestingly enough, this [idea that scientist and engineers are only one of the many acotrs] is not the reason why some respondents are quite clear about their refusal to work for the

> military, or why several other respondents said they try to steer away from controversial research. On the contrary, this notion of co-responsibility is most often put forward in those cases where the respondents feel the uses of their technologies are beneficial. Most technological designs are developed with such beneficial goals in mind, and engineers are keen to accept co-responsibility for these.

To "save" human moral responsibility, Latour's work is interpreted by some in such a manner that the "prime mover" is always a moral human agent (Smith 2003), but this does not really work (Verbeek 2008a, p. 259):

> Moralizing technology can never be the work of a 'prime mover' (cf. Smith 2003), but comes down to establishing connections between three 'agents': the designer who shapes the technology and its intended mediating role; the user who interprets and appropriates the technology; and the artifact itself, which can give rise to 'emergent' mediations, which were not intended by the designer and cannot be reduced to unexpected interpretations of users either.

It is not simply the case that Latour present a complex tool for examining ethical questions while, eventually, he will always address the human prime mover. If that was the case, we could just as well ignore the social role of technologies. Latour's explanation of the social role of technologies implies more than just a detour: initial actions are altered by human-technology interactions. Recall the metaphor of the pinball machine from the previous chapter: the initial agent is the player who is responsible for launching the ball. But after that initial act, the movement of the ball is intermediated by many different factors. The player can influence the position of the ball in certain parts of the game, but, to a large extent, the game is not steered by the player. The same holds for an agent in the Latourian actor-network: the initial agent may be a human agent, but the outcome has many different causes. Consequently, Latour's metaphysical viewpoint is problematic for promoting responsibility for the social role of technologies.

3.3.2 Distinguishing Between Different Forms of Technological Causation

Part of the solution of this problem may be found by acknowledging a distinction between human and technological causation, but it is also important to make distinctions in the human and technological types of causation. "Latour is only able to guarantee the uniformity of the culture pole [the social in so far as it consists of humans] by lumping together very distinct concepts and ideas such as power, intersubjectivity, politics, ethics, practical rationality, normativity, morality, and so on, and so forth" (Swierstra 1995, p. 26, translated). This argument can also be extended to the "pole" of technology: Latour describes all interactions between humans and technologies in terms of power relations without offering a stratified understanding of how technologies alter our actions (Harbers 1995).

One can argue that Latour uses many different concepts to address the mediating roles of different technologies—"determine", "backdrop", "authorize, allow, afford, encourage, permit, suggest, influence, block, render possible, forbid, and so on" (Latour 2005a, p. 72). Several authors have used this diversity to explain that technologies can exert their social force on different levels. Peter-Paul Verbeek has argued that artifacts can persuade, seduce or compel people to act differently (Verbeek 2008a). The first form of mediation is mild, and easy to ignore, such as a no-smoking sign. The second form is more stringent, meaning that it is harder to disregard, but still not impossible. The removal of ashtrays can serve as an example. In a hotel lobby without ashtrays one can smoke a cigar: but throwing the ash and the stub on a white carpet is rude. Enforcing technologies leave no room for undesired actions: a fire system that starts sprinkling water as soon as it detects smoke does not leave much choice to uninvited smokers.

Introducing different levels of mediation provides some room for distinguishing between different kinds of causation, but it is only fruitful for discussing relatively simple technological tools that do not require large techno-social networks like cell phones. Since it does not explain the mechanisms behind the mediation, it quickly becomes blurred when the accumulative formation of associations is taken seriously. A particular person in an actor-network is only one agent among a wide variety of compelling, seductive and persuasive technologies. How can we successfully exert agency in a networked environment where many other entities exert some kind of social force?

The problem of the many hands is already rather substantial without adding technologies to the relevant social sources. We cannot simply ignore the important role of technologies, but would it be possible to get a clearer concept of how technological mediation actually works? If we are able to get a clearer understanding of the mechanisms of technological mediation, we may be able to be more successful in steering the social roles of technologies. ANT is useful because it shows that technologies exert social force, meaning influencing actions, but at the same time, it provides little means for distinguishing between distinct causes (Lindeman 2009):

> The orthodox STS community [the STS community that adopts ANT in the strictest interpretation] has difficulties in accepting that there are different types of agents, leading to different types of causation. The idea that there are different types of causation, needs to be brought back in STS.

For obtaining a deeper understanding of how certain things came to be, we need an approach that enables us (a) to distinguish between human and technological causation and (b) to conceptualize the different forms of technological mediation. The problems of the many hands can thus be brought back to its original human proportions.

3.4 Foreseeability and Black-Boxing

The fourth criterion for taking responsibility is that one should be able to foresee the consequences of one's actions. Technological developments are hard to predict: "few people will blame Diesel for the environmental problems caused by automobiles with diesel engines" (Swierstra and Jelsma 2006, p. 313). This argument also holds for taking responsibility prospectively: one cannot meaningfully reflect on the desirability of actions and adjust them for the better if it is unclear what the consequences will be.

The social role of technology is often hard to foresee. At the beginning of the twentieth century when asbestos was widely introduced, no one imagined that it would bring serious health risks. On the contrary, adding asbestos to paint and construction materials reduced the risk of fire dramatically. People even cooked on plates of asbestos. It took some decades before it became clear that the substance caused (long) cancer. At the start of the twentieth century, who could have imagined that asbestos was carcinogenic?

Since the impacts of technologies are hard to foretell, Hans Jonas was one of the first to define what is now known as the "precautionary principle" (Jonas 1984, p. 31):

> But just this uncertainty, which threatens to make the ethical insight ineffectual for the long-range responsibility toward the future . . . has itself to be included in the ethical theory and become the cause of a new principle, which on its part can yield a not uncertain rule for decision-making. It is the rule, stated primitively, that the prophecy of doom is to be given greater heed than the prophecy of bliss.

Since appearance of this text, numerous publications have discussed the principle, often focusing on the environmental impacts of technologies. Many distinct definitions of the principle—allowing for more or less certainty and harm—have been developed, and thus, the exact meaning is unclear. But commonly, this principle states that (Turner and Hartzell 2004, p. 449): "it is better to take precautionary measures now than to deal with serious harms to the environment or human health later on." Latour reformulates this precautionary principle rather controversially (Latour 2002, p. 258):

> The precautionary principle, so much in fashion, does not simply mean that we stop taking action until we are certain about the innocuousness of a good, for that would once again return us to the ideal of mastery and knowledge by demanding certain knowledge about an innovation which, by definition and like any technology, forever escapes mastery. On the contrary, the principle of precaution resides in the permanent maintenance of the impossibility of folding – which is precisely what technology aspires to: whence the permanent conflict of ways of being.

Latour rightfully argues that we cannot stop acting, and that we need to act to move forward. To put it simply, what will happen is never sure, and we cannot stop all developments. To improve certain situations we need to act and to take risks. But though we will never know for sure what will happen, we should at least be able to make an educated guess when we are introducing new technologies into society. This can be done, according to Swierstra and Jelsma: "some respondents did feel that, in their line of research, it is quite possible to foresee how the outcomes of their work will be used: 'everyone who does not live in an ivory tower knows what use society will make of his work'" (Swierstra and Jelsma 2005, p. 208).

ANT however, implies that techno-social change is inherently unpredictable. ANT aims to offer a perspective for studying past or current human-technology interactions and does not aim to provide more means to reflect on possible future technological mediation. ANT offers an approach to study the social role of technologies, while simultaneously is stressing the fluidity of the social. One may make an analogy to a pinball machine when discussing the third starting point of ANT. Due to the many interacting elements it is almost impossible to predict where the ball will be within 20 s. Though the timetable is less dramatic, the same holds for the techno-social networks Latour describes: the social role of artifacts is hard to predict due to the many transformations that take place.

To make the unpredictability of the techno-social change explicit, Latour presents human-technology interactions as a black-box and offers ANT as an approach to study this change. ANT aims to open the black-box by tracing the associations in time. But for "tracing" there has to be a "trail" (Latour 2005a, p. 53), which implies that the changes in human-technology associations can only be studied afterward. For taking a forward-looking responsibility this is not very helpful. Latour, however, developed his approach to study the techno-social change anthropologically. His aim was not to develop a tool for scenario studies or any other forward-looking academic field that seeks to explore different desirable futures. Instead, ANT is an approach to clarify why certain changes took place in the part. In *Aramis* for instance, he explores why a technologically advanced metro system failed to become a reality (Latour 1996). Nonetheless, in order to take responsibility for the social role of technologies, we need forward-looking tools.

3.5 Responsibility: Three Problems, Three Solutions

What can be concluded from all this? When confronting the conditions for responsibility with ANT, we come across three problems:

- It is not clear how human agents can act willingly
- It is not possible to distinguish between the distinct causes
- It is hard to foresee the techno-social developments and the consequences

The first two problems are partly the result of Latour's output-oriented idea of agency. Thus ANT is solely oriented toward consequences, and ignores that people have aims, desires, and intentions on which they can reflect and that are deeply embedded in their actions. Therefore, no answer can be provided to a request for more forward-looking responsibility, since this requires people to reflect on their aims, desires and intentions. Further, little distinction between different kinds of technological causation is offered, which makes it hard to "design" technological mediation.

The third problem is not inherent to ANT: the experiences of the nineteenth and twentieth centuries have taught that techno-sociological developments are hard to foresee. Unfortunately, ANT does not provide many tools to explore the possible future developments, but then, this is not the aim of sociological anthropology.

But what does this all imply? Should we accept the lessons of ANT and thus abandon attempts to practice a forward-looking responsibility for the social role of technologies? Or should we just ignore ANT and pretend we can do ethics without taking into account the reality of complex human-technology interactions? To discuss this question, Tsjalling Swierstra and Jaap Jelsma (Swierstra and Jelsma 2005, 2006) have completed an extensive case study based on an existing assessment project and qualitative interviews. Their results show that although difficult, it is feasible to fulfill these conditions within engineering practices. At the same time, ANT has proven to be valuable in many case studies as well. Would it be possible to combine the insights of both fields?

An often-heard criticism is that Latour has little interest in normative ethics—a limitation which he is ready to admit (Latour 1995b). But unfortunately, the opposite is also true: so far mainstream normative ethics has largely ignored the important messages of ANT. The most straightforward explanation is that ethics and STS have different ideas on the aims of studying technological development and, thus, work with different concepts (Hoeyer 2006). Nonetheless, ANT is less counter-intuitive than is often argued by ethicists. When Swierstra and Jelsma asked engineers whether they could take responsibility, they concluded (Swierstra and Jelsma 2006, p. 317):

> The lesson that the SESR interviews seem to teach us is that engineers themselves point out what techno-sociologists call "the network character" of technological development. In agreement with that conception, individual engineers seem to perceive their moral agency as restricted. They experience their practical options as being severely limited by the fact that technology development is not a matter of isolated individuals but is essentially a complex, collective affair of actors with conflicting agendas. In fact, the engineers seem to feel that this structural context makes them rather helpless and powerless in exercising possible moral duties as scientists. That feeling seems to suggest indeed that the conditions for exercising moral responsibility in normal techno-science practice are not very favorable for individuals.

However, merely accepting this view is too simple, as becomes clear when studying the behavior of scientists in the real case of developing biodegradable polymers (Swierstra and Jelsma 2006, p. 222): "they [engineers] do not appear quite so helpless". Scientists play an active role in shaping social networks: they have to if they want to find partners, grants and commissioners. They "invent" technological artifacts and actively produce a "knowledge infrastructure".

This provides scientists and engineers with a special responsibility in two ways. First, they actively create the social networks necessary for a successful development of technologies. Second, the scientists and engineers have a specific responsibility since they are considered to be "the inscribers of scripts" (Verbeek 2006, p. 369):

> The phenomenon of technological mediation creates a specific responsibility for designers. The brief overview of technical mediation in the previous section in fact shows that engineering design is an inherently moral activity.

But of course, "this cannot possibly cover all emergent mediating roles of the technology". Practitioners are not the only ones who should accept a forward-looking responsibility. ANT shows that the policy makers, retailers, users, and the many other actors also play an important role. Inscribed morality does not equal the social role of the artifact, because this social role is also dependent on the social context in which it will function and its cultural interpretation. Technologies are merely messengers of morality. The interaction between humans and nonhumans leads to "the creation of a new goal that corresponds to neither agent's program of action" (Latour 1999b, p. 178).

The Internet is the most commonly mentioned example on how the maker-user distinction becomes blurred: users use the internet to search for information, and often simultaneously, produce the internet. But when studying socio-technological change, this argument seems to hold for most (if not all) technologies. Therefore, we should not only focus on the people who invent the technologies in the laboratories when we discuss responsibility issues: also other practitioners involved in the technological development such as managers, commissioners, NGO's, and so on, need to be addressed.

So, the question is: can we ask these different practitioners to accept responsibility if we offer them a social theory that does not empower responsibility? ANT teaches that (a) the autonomy of people is severely limited, (b) technological artifacts fulfill an active social role, and; (c) the development of the techno-social networks is hard, if not impossible to predict. In promoting people to accept responsibility we need a social theory that empowers people to overcome these obstacles, at least to a minimum extent.

As Latour notes, "theories of what the social is or should become have played an enormous role in helping actors to define where they stand, who they are, whom they should take into account, how they should justify themselves, and to which sort of forces they are allowed to bend" (Latour 2005a, pp. 230–231). In taking responsibility it is important that people think of themselves as persons who can exert moral agency: in being prepared to accept responsibility, people should have the idea that they can make a difference, however small. As long as people adopt a social theory that teaches them that they are only one small, replaceable cog in a machine (or node in a network), they are not likely to become proactive. What is the use of sticking out your neck if it does not make much difference anyway (Swierstra 1995)?

If we want practitioners to take a forward-looking responsibility, we have to provide them with a social view that empowers. Yet, such a view should do justice to the social role of technological artifacts. We cannot simply ignore Latour's findings just because they do not suit the political aim to get people to take responsibility. So the question is, how can we bring together mediation theory with responsibility issues? For this, we need three solutions:

1. We need an understanding of human agency, which recognizes we are part of larger socio-technological networks, and of our selves as moral agents with multiple reasons of action. The first challenge is to see whether we can develop an understanding of human moral agency based on their reasons of action, without presuming to be free, atomist and rational human beings.
2. We need to conceptualize the various forms of mediation, to better understand how technologies alter human action. The second challenge is to explore whether

we can reformulate Latour's mediation theory in a manner that pays respect to the experience of the different kinds of causation.

3. We need some practical tools to open the black-box prospectively. The third challenge is to see whether the important insights of ANT can be reinterpreted for exploring future developments.

The following chapter deals with the first challenge by exploring whether Alasdair MacIntyre's analysis of human agency can serve as an alternative understanding of human agency that meets our needs. His practice-based virtue or life ethics recognizes the social and biological surrounding of people, which makes his approach less vulnerable to Latour's criticism that ethics assumes free, atomist human agents. At the same time, MacIntyre offers an understanding of agency that is not solely output-oriented but considers aims, desires and intentions.

Chapters 5 and 6 then deal respectively with challenges two and three, by offering a reformulation of mediation theory that leaves more room for different kinds of causation and by developing an empirically informed approach for exploring future developments.

Chapter 4
Becoming Responsible for Techno-Social Practices

At the end of the previous chapter, we were left with three challenges:

- The search for a networked understanding of responsible human agency
- The search for a deeper understanding of how technologies mediate actions
- The search for some tools to help us to see the possible social roles of technologies

This chapter picks up the first challenge. It aims to develop an integrated anthropological and ethical position that recognizes people as living and acting in techno-social networks. Simultaneously, the ethical position aims to empower people to take responsibility for their own roles in these techno-social networks. In this way, it becomes feasible to take a forward-looking responsibility for the social role of technologies, because the essential aspects of ethical theory are brought back into ANT. To put it the other way around, a forward-looking responsibility for the social impact of technologies will be made feasible by incorporating the lessons of ANT in ethics.

4.1 Different Styles of Doing Ethics

Thus, the aim of this chapter is to present an understanding of human agency that (a) considers the embeddedness of people's actions in techno-social networks, and (b) takes our ability to reflect on our reasons for actions such as moral beliefs, aims, desires and intentions seriously. As indicated in Chap. 3, this aim is controversial. Bruno Latour has developed ANT for tracing normativity in techno-social networks while rejecting what he sees as the most prominent modernist ethical theories (meaning utilitarianism and deontology). His classical training in philosophy makes Latour aware of at least three closely related drawbacks of these mainstream ethical approaches (Latour 1995b).

First, he argues that the distinction made by these ethical theories between exerting power (putting a gun to someone's head) and providing reasons (arguments that

K. Waelbers, *Doing Good with Technologies: Taking Responsibility for the Social Role of Emerging Technologies*, Philosophy of Engineering and Technology 4,
DOI 10.1007/978-94-007-1640-7_4,

someone needs to act differently) is false. According to Latour, arguments should also be understood as forces and all forces are normative in the sense of being prescriptive for action. He argues that because these rationalist ethical theories accept this false distinction, they do not recognize morality in science and technology. Science and engineering were believed to be about forces, while ethics was believed to be about moral values. However, in ANT forces and reasons cannot be so easily distinguished. We have already seen in the previous chapter that technologies mediate human action. Furthermore, persuading someone with arguments is a form of exerting social power (see also Latour 1992). There is no principled distinction between providing reasons and exerting power. This belief is at the core of Latour's argument that technologies are moralizing. An ethical theory that presumes a dichotomy between exerting reasons and explicating power is not reconcilable with Latour's ANT.

Second, Latour believes that we should study moral beliefs and factual ideas in the same manner. Both are the result of negotiations that take place in networks. There are no objective scientific facts and there is no objective moral truth, instead both "facts" and "morals" are established by negotiations in techno-social networks, (see also Latour 1987b).

Third—and at this point most importantly—ethical theories should not be about "giving an account of one's actions" (Latour 1995b, p. 35): ethics for Latour should be about "sorting different possibilities" or choosing among different options. What does this mean? Latour argues that since all "possible worlds" (or possible futures) cannot exist simultaneously, we need to choose. Morality arises in the attempt to shape the world. According to Latour, morality has not so much to do with searching for an ethical foundation: he argues that ethics is about acting in the world to create a possible future that is desired. People will not alter their actions because ethicists find, or fail to find, an objective foundation for moral truth. For Latour, morality arises in performing, acting, choosing and deciding about the possible futures. In these acts, not just arguments, but "many elements together constitute the virtuous" (Latour 1995b, p. 35). "Elements" are humans, technologies, actions, associations, conversations, and so on.

These three objections, however, result from a narrow understanding of what ethicist do: Latour understands ethics as an academic discipline like epistemology (Swierstra 2000). Epistemology is a rigid research focused on the question of how to obtain objective scientific knowledge. Similarly, Latour imagines the work of ethicists as the search for objective moral knowledge, and he points out the non-utility of such an exercise. In everyday life, norms and values do not come about by a rational and methodic search for foundations of good, right or just. Instead, Latour explains, morality is brought about by interacting agents in networks of power.

But, this narrow interpretation of what ethicists do, ignores moral theories that adopt a broader focus. Several schools of ethics see people embedded in social structures that form and guide them. Such theories acknowledge a classical, broader integrated view that includes practices, politics, philosophical anthropology, and ideas on a good life. Consequently, they present a more context sensitive and realistic account of what ethics is and can do. This more context sensitive approach is often identified with virtue theory, but a better term would be life ethics, since virtue ethics suggest that this school of ethics only focuses on having a good character while good life theories actually include broad views on humanity, morality (norms, values and virtues), actions, friendships, happiness, capabilities, politics, society and social practices, etc.

Life ethics is not susceptible to the three objections Latour expressed. Latour's arguments mainly address classical notions of deontology and utilitarianism. Life ethics focuses on how to live and have a good life, without assuming objective moral rules and without assuming autonomous, atomist subjects. Instead, these theories focus on how to be a moral person who has to act in real-life situations. Thus these theories incorporate the social context, the social forces, the culture, and the interests, characters and lives of the people involved.

Next to life-ethics, there are other schools of moral philosophy that reject the decontextualised, atomist agent and the objectivity of moral principles or methods of deontology and utilitarianism. Contemporary life ethics however, is the most promising since it offers an elaborate understanding of what it means to be a human agent, in relation to what it means to be and do good. Furthermore, there is a tradition in life ethics that connects the ethics of an individual to the social. Casuistic ethics (which is pre-eminently an ethics acknowledging that what is good to do depends on the specific situation of the case) and hermeneutical ethics for instance, all recognize that the only reasonable manner in which to do ethics is a contextualized manner, but they do not offer such a comprehensive worldview and anthropology as many virtue ethicists do. One of the main reasons for this is that virtue-ethics has a 2,500 year old tradition in which many people have worked to complete it and make it as coherent as possible.

Life ethics has not been popular for quite some time: it did not comply with the desire to find an objective method for proving moral principles that many moral philosophers inherited from Kant, Mill and Bentham. However, this alternative form of understanding human agency regained more attention in the 1980s (Pence 1991). Anscombe (1981/1958) reintroduced life ethics by explicating the missing authority of utilitarianism and deontology. She argued that in contemporary Western societies, there is no God, natural law, or rational inevitability commanding us how to act. Consequently no objective foundation for ethical rules can be given. We were, Anscombe argued, in need of an alternative, and this alternative can be found in life ethics.

Following Anscombe, several authors—such as Martha Nussbaum (1986), Charles Taylor (1989) and Alasdair MacIntyre (1985)—have developed influential good life theories. For the purpose of this study, this chapter places particularly interest on MacIntyre's practice oriented ethics. First, because his practice oriented approach can serve as an addition to Latour's network approach in such a way that mediation theory leaves more room for studying moral aims, desires and intentions. Second, MacIntyre's anthropology provides an understanding of how the morality of people is formed in a socially embedded manner. Third, MacIntyre's understanding of practical reasoning explains how we can regain some responsibility, even though we are not autonomous agents who can control their own course of action.

The next three sections of this chapter discuss these three points. But before discussing MacIntyre's life ethics, it is important to explain his analysis of theories such as utilitarianism and deontology. This provides a deeper understanding of how his view differs from these enlightenment-based ethical theories, and explains why his theory is less susceptible to ANT.

4.1.1 Why Objectifying Ethics Failed

MacIntyre explains that deontology and utilitarianism had to fail because: "its key premises would characterize some feature or features of human nature; and the rules of morality would then be explained and justified as being those rules which a being possessing just such a human nature could be expected to accept" (MacIntyre 1985, p. 52). However, this conception of human nature conflicts with its aim to develop universal moral principles. On the one hand, people are urged to live an autonomous life, meaning that they have to decide their own moral rules or values rationally. On the other hand, these rules are considered to be universal, meaning that the rules should be acceptable for, and applicable to all persons in comparable situations. This obviously creates a tension, and, in line with this dilemma, the two premises did not provide a coherent and accurate set of rules. Therefore, contemporary western societies did not manage to create one moral culture in which all people rationally accept a substantial set of shared moral rules. Instead, many different subcultures can be recognized, each with their own morals.

According to MacIntyre, the lack of a substantial set of commonly shared moral believes is the result of the failed attempt of the deontology project to "provide for the debate in the public realm standards and methods of rational justification" (MacIntyre 1988, p. 6). Further liberalism argued that ideas about the good life belong to the private domain and governments should not meddle in people's private lives (Waelbers and Briggle 2010). MacIntyre explains the inconsistency between the two foundations. The first foundation is the belief in human autonomy—the power to impose objective, rational, universal, moral laws on oneself. The essence of human nature is autonomy, and exerting this self-determination is the most basic moral value in deontology. Humans are understood as free beings, who are (in principle) able to reflect before acting. The autonomous agent decides which action is the most moral based on self-imposed, universal rules or on the assessment of the possible outcomes.

Second, according to these ethical theories, the rules or assessments the autonomous agents formulate are morally valid if and only if they are universal—meaning that they have the same moral force for all agents in comparable situations. Expressed in Kantian terms: "act only according to that maxim whereby you can at the same time will that it should become a universal law" (Kant 1785/1993, p. 30). Also utilitarianism and consequentialism—theories in which the results of the actions have a more central place than the moral rules—presume that their way to determine which action is the most desirable is highly universal, though the individual agents should have the freedom and rationality to determine their own actions (MacIntyre 1988, p. 3):

> Rationality requires, so it has been argued by a number of academic philosophers, that we first divest ourselves of allegiance to any one of the contending theories and that we abstract ourselves from all those particularities of social relationship in terms of which we have been accustomed to understand our responsibilities and our interests. Only by doing so, it has been suggested, shall we arrive at a genuinely neutral, impartial, and, in this way, universal point of view, freed from the partisanship and the partiality and one-sidedness that otherwise affect us.

Contemporary, liberal societies are so familiar with these two foundations that, in daily life they are rarely contested. Our moral right to freedom of religion, opinion and speech are highly valued. However, MacIntyre argues that on closer inspection, it is not that hard to see that these two premises create problems. The first premise tells that we need to be the author of our own moral rules (we should determine freely and rationally what we believe, do and say), while the second states that those moral rules are universal (I should will that everyone does what I do, or everyone should act in such a way that it maximizes happiness). Thus, MacIntyre explains, in our contemporary society—in which people are reasonably free to determine their own moral views—they come up with different sets of moral viewpoints. These different sets are often conflicting, and not accepted as being universally applicable (MacIntyre 1988).

MacIntyre explains that this variety in moral beliefs makes the question "How to live a good life?" difficult to discuss. MacIntyre does not pay much attention to technologies, but in contemporary high-technological cultures this question is even more difficult to discuss (Waelbers and Briggle 2010). Technologies are implemented on a public level, but they have substantial impacts on the lives of individuals. Think for instance about biotechnological approaches for crop improvement and individual desires to eat gen-tech free food. The new media which influence everyone's live and the perception of what a good life is. How can we deal responsibly with such new technologies? The study of public debates teaches that people strongly disagree on these kinds of questions. We do not share a common method or a set of common moral principles that can settle the debate.

MacIntyre sees this lack of communality in morality as the contemporary social crisis: a deep crisis in our moral debates that arises from the wrongful presumption of deontology and utilitarianism that ethics could be objectified or rational. MacIntyre shows that in public (or political) debates on issues such as abortion, global warming or privacy-violating information technology, debaters use different moral arguments. These arguments are based on different moral theories that have their own historical and cultural contexts. In debates, people do not give an account for how the arguments relate. In public debates, participants use human rights, facts of nature, moral values, and emotions to back up conflicting viewpoints, without being able to clarify the interrelations and without referring to a common background or world-view. As an example, MacIntyre refers to the debate on abortion. Arguments on the rights of the mother and the prevention of an unwelcome and unhappy child, oppose those arguments where ending an innocent life is wrong and that you cannot will that your mother had aborted you. We can argue back from these kinds of arguments to the premises, but still, we end in disagreement with rival premises. There is no generally accepted method or coordinating theory to discuss moral issues.

This difficulty becomes particularly pressing when discussing moral issues on new technologies, since (1) we lack experience with dealing with these technologies on which we can base our judgments, (2) we lack established moral viewpoints on the new possibilities these technologies bring, and (3) we cannot foresee what the impacts of the technologies will be.

There is much disagreement about almost all arguments in the public debates, since a shared argumentative structure behind the positions is lacking. The same debaters use arguments that are derived from different perspectives, practices and traditions. Therefore we *seem* to live in an "emotivist culture" and ethics is in "crisis" (MacIntyre 1985). MacIntyre's point is not that we should all agree with each other and that moral disagreement is bad in itself: he criticizes the form of many contemporary debates on a deeper level. Different moral principles and guidelines are used in the same disputes, without any reference to how they relate to each other. As a consequence, moral statements seem to be merely utterances of emotions or preferences and the arguments cannot be balanced. Moral debates rarely end in consensus. They become frustrating, since what we feel as important moral insights become, due to this use of incomplete moral language, moral expressions that are regarded as utterances of like or dislike (MacIntyre 1985). Moral arguments lose their social strength, since they "have come to be understood…as weapons, the techniques for deploying which furnish a key part of the professional skills of lawyers, academics, economics, and journalists who thereby dominate the dialectically unfluent and inarticulate" (MacIntyre 1988, p. 133). Note that this observation resembles Latour's view that providing moral arguments is similar to exerting social power.

However, MacIntyre's analysis does not stop here. Although, in the beginning of *After Virtue*, MacIntyre sketches the moral crisis of our debates, he recognizes that "when someone utters a moral judgment such as 'this is right' or 'this is good', it does not mean the same as 'I approve of this; do so as well' or 'hurrah for this!' or any of the other attempts at equivalence suggested by emotive theorists" (MacIntyre 1985, p. 13). Moral arguments go deeper than just uttering taste or preference. Understanding the fuller nature of morality reinstates the motivational power and authority of moral arguments. Additionally, even though much disagreement exists in political debates, in daily life people commonly get along quite smoothly. We are not constantly debating our moral beliefs because we commonly share enough moral views with those whom we work and live. These shared moral beliefs also have more social power as Latour recognizes. How can we explain this?

MacIntyre answers by explaining that in the practices people can act based on shared intrinsic motives. Consider what this might mean and how such a view can serve as a valuable addition to ANT.

4.2 Intrinsic Motives and Practices in Actor-Networks

Even though people disagree on many moral principles, we are able to participate cooperatively in practices, MacIntyre explains. This understanding is valuable for my aim to introduce ethics and morality in ANT. The concept of practices can serve as a valuable addition to ANT: practices are part of the networks Latour describes and enable people to recognize intrinsic motives such as virtues, moral beliefs, and intentions. Practices are (MacIntyre 1985, p. 187):

> any coherent and complex form of socially established cooperative human activity through which goods internal to that form of activity are realized in the course of trying to achieve those standards of excellence which are appropriate to and partially definitive of, that form of activity, with the result that human powers to achieve excellence, and human conception of the ends and goods involved, are systematically extended.

With this definition, MacIntyre's introduces the following line of thought: to obtain the products of the practices, people have to cooperate in those practices. In order to be successful in obtaining those goods, people learn how to be their better selves by cooperating with others. Consequently, they acquire some virtues—characteristics needed for moral communication and moral actions such as right ambition, patience, truthfulness, wittiness, friendliness and creativity.

Philosophers use the term virtues, but non-philosophers would use the word "character" (Pence 1991) or "identity". These words have a less antiquated connotation. Being antiquated is not a problem because using old terms is not fashionable, but the meaning of the word has changed radically over the last three millenniums. Furthermore, the words "character" and "identity" have a broader, less idealistic connotation, making it more suitable for addressing real-life situations.

In the practices, common moral grounds for people can be found, and in the practices, it becomes clear what it means to be responsible. Practices are the frameworks in which we develop our moral customs and, these provide a normative perspective that forms an important source for reflection on moral actions. In other words, practices form a source for taking responsibility. Elaboration on this point can proceed by elaborating on three levels of describing practices. Practices are (1) kinds of activities, that are (2) teleological (they are directed at the production of intrinsic goods) and (3) intrinsically moral and moralizing.

Ad 1. Practices are kinds of activities in MacIntyre's work, practices are (1) coherent, complex and (2) socially established actions that (3) are carried out through human cooperation, and (4) involve technical skills (Cooper 1987, p. 321). First, practices exhibit some coherence and complexity. Throwing a ball against a wall is not a practice, but playing basketball is. Second, the rules of this practice are socially established, meaning that are generally recognized. The rules of basketball are almost the same all over the world. Third, practices are carried out through human cooperation; one cannot organize a basketball competition alone, nor can one develop technologies without working in a team. Fourth, technical skills are needed in order to succeed in the practices.

But there is more to practices than these four elements. The main characteristic of practices is that they are teleological in the sense that they are oriented toward some good. Therefore they are intrinsically moral as well as moralizing.

Ad 2. Practices are teleological Practices such as creating art, practicing medicine, caring for children and conducting scientific research are established for a reason: people want to achieve predefined intrinsic goods such as realizing a beautiful painting or musical performance, health, well-raised adolescents and knowledge. Internal goods are distinct from external goods. External goods are personal properties and possessions. Since they are scarce, external goods are objects of competition. If one gains external goods such as power, money, property or fame, others lose some. Internal goods can also be achieved in competition, but their achievement is good for everyone in the practice: internal goods are not scarce and someone's possession of an internal good may benefit many. If someone excels in a certain practice, the skills or techniques of that practice can be enhanced. For instance, becoming an excellent oncologist performing brilliant research and providing outstanding patient care adds to the development of the entire practice of oncology.

Ad 3. Practices are intrinsically moral and moralizing Practices are intrinsically moral since they are directed at internal goods: these goods are needed for a flourishing human existence. Put differently, internal goods such as art, knowledge and health help sustain a good life and can only be obtained by virtuous cooperation. Therefore practices are intrinsically moral activities. Taking part in practices is also moralizing since in striving for internal goods people strive for excellence, and (MacIntyre 1998, p. 240):

> the achievement of excellence in activity characteristically requires the acquisition of skills. Furthermore, without virtues and good customs, skills lack the directions that their exercise requires, particularly if excellence is to be achieved. So, it is characteristic of such practices that engaging in them provides a practical education into the virtues.

Engaging in practices increases human capacities to achieve internal goods, and enables people to acquire virtues. "A virtue is an acquired human quality the possession and exercise of which tends to enable us to achieve those goods which are internal to practices and the lack of which effectively prevents us from achieving any such goods" (MacIntyre 1985, p. 191). MacIntyre explains that a large part of our morality finds its origin in practices: by cooperating with others we learn to distinguish better from worse. But people are not just morally formed by practices: people also shape the morality of practices. For example, an oncologist is trained in a pre-existing medical practice, and she is improving the practice at the same time when she is performing well.

4.2.1 Practices Versus Networks

How can a practice oriented ethics be a useful addition to ANT? MacIntyre's ethical theory is not primarily about "giving an account of one's actions", which Latour dislikes so much in ethics (Latour 1995b, p. 35). Instead, with MacIntyre's ethics we can start wondering what it means to be and to do good in a network of practices. This suggests that a MacIntyrean ethics can support an ethical outlook in ANT. Although MacIntyre has not written much on technology, it is not unimaginable that his philosophy can enrich ANT. His life ethics allows for technologies to play an important social role (or, the other way around: the important lessons of ANT can be incorporated in a practice oriented ethics).

At the conference held in Dublin to celebrate his 80th anniversary (MacIntyre 2009b), MacIntyre gave some shorts remarks on the suggestion that technologies such as computers, mobile phones and the internet influence our societies. In an extensive answer, he said among other things the following (quoted with permission):

> Aesthetically and in many other ways, I think computers are amongst the most extraordinary inventions, but what they affect of course, and this is what the internet is about, is information flow. The way in which information flows, the extent to which information is shared, the way in which information in shaped, and how people respond to information is enormously important. It is very clear that the things that happen through computing, the internet and text messaging are quite different: lines of communication are opening up. And I take it to be a fact that nobody has reflected enough, yet, on what the implications of this are. If you compare this with the last great change, and the last great change was television, what could not be predicted was the time that people spend watching and how this changed society. Look at the reports of the Norwegian Democracy Project. People in Norway stopped going out in the evenings to meetings, and different kinds of community organizations disappeared. Those organizations turned out to be of crucial importance for mediating between citizens and larger political organizations of local and national governance. So actually, the television set altered Norwegian politics in various ways, leaving most Norwegian big parties now pretty much in doubt about their constituency.

In the second part of his answer, MacIntyre explained how he thought the Internet supports education and entertainment. But he also added that people may get an inaccurate feeling of being in touch with what happens in the world, since many Internet sources—if accurate at all—lack the important background information that for instance quality newspapers supply. Shared knowledge about the backgrounds of political events or news items is decreasing, according to MacIntyre, since people use a plurality of Internet sources as an information base. As a result they lack a shared ground for judging the news.

These loose reflections by MacIntyre about the social roles of technologies do not seem to conflict with ANT and his practice oriented ethics does not seem susceptible to Latour's criticisms against ethics. But, the huge differences between both authors cannot be denied. MacIntyre focuses on moral motives (or reasons), while Latour addresses techno-social forces. ANT's intention is to analyze the outcomes of associations of actants in order to explain social change, while a practice oriented ethics explores how our virtues, moral viewpoints and moral beliefs are formed.

The difference in approach between Latour and MacIntyre can partially be explained by the distinct research questions of the two fields. ANT is developed as a methodology to study sociologically (or anthropologically) human-technology relations and changes, while the aim of MacIntyre's life ethics is to describe how we can become morally good people. Since the aims are different, the approaches are distinct, but this is not an intractable conflict.

The stronger points of ANT can supplement the weaker points of MacIntyre's ethics and vice versa. MacIntyre solely focuses on virtues and practices while many human activities are not practices in his sense of the word, and human programs of actions (to use a Latourian term) are not only formed by their virtues. In addition, MacIntyre does not recognize the social role of technologies. Latour's ANT points out the social role of technologies, and he covers a much broader palette of human actions, but his externalist manner of studying the networks is blind to the moral intentions and aims.

The strong points of both authors can be brought together when noticing that they are both discussing interactions. Latour argues that morality is the result of the interaction between human and technological agents, and MacIntyre argues that the morality of a person is the result of interactions in practices. To get a more comprehensive understanding, it is reasonable to take practices as parts of the techno-social networks and see morality as the result of interaction in networked practices. Put differently, practices are small networks that are part of larger systems in which humans and technologies deeply influence each other. Take the practice of engineering as an example: such a practice is part of a larger network. Universities aim to improve knowledge, companies aim to make a profit by launching new products on the market, users want to purchase new high-tech goods, and governments are working to regulate market releases.

A rare yet valuable remark MacIntyre makes about technologies, is that engineering itself is a practice (MacIntyre 1985). This remark is important since it implies that MacIntyre considers engineering to be an ethical endeavor in itself. In other words, engineers (and those who are closely cooperating with engineers) need a shared morality to function well. Engineers need shared intrinsic motives for their work. A similar argument can be made for other actors involved in technological change. Parents apply technologies to take better care of their children, artists employ new media to express themselves, and physicians employ technologies to be able to diagnose and cure their patients. All these groups contribute to technological change, while their practices are also driven by intrinsic motives.

Seeing practices as being parts of techno-social networks is helpful to recognize that it is not enough to describe society only in terms of human and technological actors that exert social forces. The notion of practices reminds us of the (moral) motives behind actions. This chapter focuses on the advantages of the concept of practices additional to the concept of networks.

For understanding the techno-social sphere—which is the aim of ANT—MacIntyre's concept of practices provides an additional dimension. But the concept of practices can also help bridge the gap between Latourian approaches focusing on the context that shapes human action, and ethical theories that focus on the autonomous human agency. In order to participate successfully in practices, one has to internalize a shared morality and learn how to act in a desirable ways. Therefore, people are not atomist subjects, which does not imply that they are determined by practices: this can be compared to Latour's networks that are established by people, while they also shape the actions. Practices change constantly as they are constituted by people, who still have the capacity to reflect on their morality and their actions and change them correspondingly. Humans give shape to the morality of a practice—the result of the interactions of the others involved in the same complex activity—while their morality is simultaneously shaped by the morality of the practice. However, this mutual construction in the practices is not a conservative process, but a fluid one. This fluidity is emphasized by the idea that people develop themselves not just within one practice; they participate in other practices and activities as well and are able to reflect on these different experiences and form new views. With these new views they can enrich existing practices and in this way a dynamic system of feedback arises.

Understanding practices as being part of the networks makes it feasible to study the moral reasons behind the actions and provides a counter argument to Latour's view that explicating moral arguments do not motivate people to act differently. Just like Latour, MacIntyre recognizes that deontological and utilitarian ethical theories need to deal with the problem of motivation (MacIntyre 1985). MacIntyre argues that being moral is essential for participating in practices, and the presumption is that many people want to be good in at least some practices (for instance the practice of parenting). We want to be good parents, good athletes, and good physicians. In all such cases, we strive to be moral.

Additionally, in our daily existence in high-technological cultures, we need virtues for living and having a good life. MacIntyre asks himself: "what would a human being lack who lacked the virtues?" (MacIntyre 1985, p. 201). First, persons who lacks virtues will not succeed as participants in practices, for they will be involved in too many conflicts and too much arbitrariness. Second, the life of a person who lacks virtues can be defective as it would not resemble the good kind of life a man or woman wants to live. Although in *After Virtue* MacIntyre hesitates to judge the life of a person as a whole (as he is afraid of overestimating the unity within one's life), still the argument stands that at certain time one's life can be considered poor because of a lack of virtuous. To begin with, someone with little virtues is not likely to have many friends.

Thus we need virtues to flourish as individuals, but we also need them to cooperate successfully. We cannot successfully work together in practices if we are not reliable, fair, friendly, creative, communicative or humorous. We can distinguish many well-operating practices in our high-tech societies: "every human society is marked by the presence of a multiplicity of practices, and individual human lives are characteristically marked by participation in a number of practices" (Murphy 2003, p. 163). Following MacIntyre's line of thought, this implies that most people were able to achieve the goods internal to the practices. This means that many people have adapted the practice common view on what is virtuous and share—at least to a minimal extent—a common morality that of cooperating in the practices. But of course, it is not possible to establish a society solely on virtues. Not everyone will strive for virtuous excellence and even if we all do our best, we need other forms of normativity as well. How do we decide who has to give way and who may drive at a busy crossroads, if we do not employ traffic rules and traffic lights? This is where Latour's techno-social forces offer helpful explanations.

In summary, MacIntyre's concept of practice is compatible with the lessons of ANT about the social role of technologies. However, in understanding human agency on a more fundamental level MacIntyre's and Latour's views diverge.

4.3 Embedded Agency

What does it mean to be a responsible agent in a world of practices that are part of techno-social networks? Latour complains that the Kantian concept of agency wrongfully presumes a faceless agent, with no history, cultural background, social dependence and personal circumstances. Like Latour, MacIntyre fiercely argues against this view but takes a different turn. He does not address our embedding in the material context, though he discusses extensively our biological vulnerabilities and social interdependence (MacIntyre 1999).

In *After Virtue* (1985), MacIntyre explicitly distances himself from any argument that could lead to something that looks like an Aristotelian metaphysics in which the biological features of humans are translated into visions of a good life. However, in later works, he admits that we need to address our biological condition for understanding who we are and how we can do good. In *Dependent Rational Animals* (1999), he stresses two related sets of facts, that undermine the rationalist ideal of autonomy. First are those concerning our biological vulnerabilities and afflictions. These vulnerabilities and afflictions are directly connected to our animalism. Since we belong to the kingdom of animals, we are vulnerable and suffer from hunger, disease, pain, and so on. Here, it is not very interesting whether MacIntyre's analysis regarding nonhuman animals is accurate, and thus this texts does not go into detail regarding his description of the capacities or moral status of nonhuman animals. For the purpose of this study, his view that humans are distinguished among animals and not distinguished from animals is taken for granted. The discussion focuses on the question of whether MacIntyre succeeds in explaining the full meaning of biological vulnerability and social independence, while recognizing their capacity for responsible moral agency.

Second, MacIntyre addresses the extent of our social interdependence. Dependence stems from our vulnerabilities and afflictions. It is the result of human limitation and refers to the consequent need of cooperation with others.

MacIntyre argues that in contemporary dominant ethical theories, the full extent of these two key aspects of human action is often neglected. MacIntyre thus seems to agree with Latour, since MacIntyre argues that in philosophy and ethics, human agents are wrongfully assumed to be free, autonomous beings who find themselves "continuously rational, healthy and untroubled" (MacIntyre 1999, p. 2) in a-historical, context-impoverished situations. Of course, dependence on others is acknowledged by most authors in the sense that we need other people to achieve our goals, and some—mostly feminist—authors have paid attention to the relevance of our bodies. "They have underestimated the importance of the fact that our bodies are animal bodies with the identity and continuities of animal bodies, and they have failed to recognize adequately that in this present life it is true of us that we do not merely have, but are our bodies" (MacIntyre 1999, p. 6).

The concepts "agent", "I" and "my body" can rarely, if ever, be substituted for each other without change of sense (MacIntyre 2006b). Saying "I did it" implies for instance some accountability, while saying "my body did it" does not. Though, what is true for your body is also true for you: if your body is slim, you are slim; if your body is in pain, you are in pain. Further, in our present human condition, there is no such thing as a disembodied human mind. In our life, our body and our mind coincide (MacIntyre 2006b, p. 96):

> When I pick some object and hand it to you, there are not two things going on, the one to be characterized as the intentional act of giving something to somebody, the other as a set of changes in the muscles and nerves of the arm and the hand. There is one act or sequence of acts, one change or sequence of changes. The giving is the movements of arm and hand.

But being a body does not mean that we are biologically determined. We can still take responsibility for our actions. MacIntyre's understanding of what it means to be a responsible human agent departs from the notion that people can reflect on their reasons (MacIntyre 1999, p. 56):

> It is because any exercise of the power to reflect on our reasons for action presupposes that we already have such reasons about which we can reflect, prior to our reflection. And for us human beings it is because we do have reasons for action prior to any reflection, ..., that we have an initial matter for reflection, a starting point for that transition to rationality which a mastery of some of the complexities of language use can provide.

A reason in a MacIntyrean sense is a presumption that an action leads to some aimed at good. Being cold and wanting to be warm is a reason for turning on the central heating system. Reasons can be unconscious and unexplicated, as also animals like dolphins can have reasons, MacIntyre explains. But becoming a moral, responsible being means becoming able to reflect upon one's reasons (MacIntyre 2006b, p. 96):

> It is true of me that I am accountable for my actions and my omissions, that I am aware of the objects that I perceive, that I am able to reflect not only on my actions and my experiences, but also on my reflections on my actions and my experiences.

During infancy, people should learn the ability to reflect as a moral agent. We are taught that we should have good reasons to act in a certain way. Instead of asking "what do I want", we have to learn to reflect on "What is best to do?" MacIntyre explains that even though we are biologically vulnerable and socially interdependent, we can grow toward rationality by moving from having reasons for acting in a certain way toward a "specifically human state of being able to evaluate those reasons, to revise them or to abandon them and replace them with others" (MacIntyre 1999, p. 91). In other words, we are able to learn how to take responsibility.

Of course, although we can work to be reasonably "rational", we remain dependent on other people. Social relationships such as friendship and collegiality are essential in many respects: most of our knowledge and interpretations are based on what other people tell us. Also, self-knowledge depends to a substantial degree on the feedback of others.

So, MacIntyre's philosophical anthropology (in other words, his understanding of what it means to be a responsible human agent) grants that we have to live and act in a real world. Being vulnerable and dependent of others shapes our actions on a deep level. If one adds to MacIntyre's account that people are also technologically mediated, a moral view arises that is less susceptible to Latour's criticisms against the wrongful presumptions of ethics on autonomous agency. However, when discussing responsibility questions, we still have to deal with the problem of acting according to our free will. If we are biologically vulnerable, socially interdependent and technologically mediated, how can we ever make an action our own? Can we take responsibility if our doings are also shaped by so many external factors?

MacIntyre provides an answer by distinguishing between having reasons from the activity of practical reasoning. Being able to reason—in the sense of being able to carry out everyday practical reflections about questions what to do—is a necessary condition for being a responsible person. People can evaluate the reasons for actions that are biologically or socially given by standing back from them and thinking critically. Based on this ability we can adjust our actions. This does not imply that we are atomist, rational beings: it simply implies that we can ask ourselves the question whether it is good to act in a certain way.

4.4 Explaining Forward-Looking Responsibility

What does it mean to accept and take a forward-looking responsibility in a practice oriented life ethics? Here, the following definitions are adopted: accepting responsibility means recognizing that your actions can make a future difference (no matter how local) and that you are willing to adjust your actions for the better. Taking responsibility means actually taking the effort to find out what is a good thing to do and acting according to those findings. This definition raises two questions:

- What does "good" mean?
- What does it mean to "take the effort to find out"?

The following subsections respectively discuss these two questions.

4.4.1 Doing Good

Taking a forward-looking responsibility also stands for doing good with regard to the future: but what does it mean to do good in a practice oriented life ethics? In *Dependent Rational Animals* (1999), MacIntyre distinguishes four different types of "good" that are all connected to human and environmental flourishing (MacIntyre 1999). First, "good" can refer to the care of our biological vulnerabilities. To flourish, we need to fulfill our bodily needs: we need food, sleep, cleaning and freedom from pain. Fulfilling these primary needs may provide pleasure. Note that many technologies aim to facilitate acquiring such goods: since we are biologically vulnerable, we need technologies to suit our bodily needs (like agricultural technologies, cooking devices, architecture and medication).

This perception of good is not recognized by MacIntyre in *After Virtue* since it leads to the risk of adapting an Aristotelian metaphysics (which is disadvantageous to children, women, mentally and physically challenged persons and other races, and it additionally argues in favor of slavery). In his later work, MacIntyre realized that it is impossible to ignore our physics, as we cannot experience a good life when starving and the need to cooperate with others in the practices is given by being biologically vulnerable. In addition, acknowledging our biological background does not automatically imply that one also adopts Aristotle's discriminating anthropology. On the contrary, healthy, rich men also experience bodily needs, and physical afflictions.

Second, "good" can be employed for describing the means for another goal. For instance, cars are good for transportation, but useless for hammering a nail in the wall. For humans, this sense of good refers to being good at a certain task. Being a good engineer means that you are able to design safe and efficient technological devices. In this sense, one can also be a good thief or assassin. But of course, being a thief or assassin does not equal being a good person. Therefore MacIntyre distinguishes two other meanings of good in relation to human and environmental flourishing.

Third, someone can be said to be "good" when he or she fulfills a socially desirable function or role. This kind of good is taught to adolescents when they train for vocations or professions. In this interpretation, actions are good insofar as they support the genuine goods internal to that activity. Genuine goods (or internal goods, see also Sect. 4.2) are ends worth pursuing also for their own sake, such as farming, artistry, parenthood. This type of good refers to being "excellent in achieving the goods of this or that particular activity", which is socially valued. However, one can be good in this sense, but not be a good person. For instance, one can be a prominent ethicist and a pedophile at the same time.

Finally, being "good" can mean being virtuous. Here, the question is not what is good to do or to have as an agent engaged in a certain activity or role, but what being a good person entails. This question is put in a broader perspective, transcending the level of the particular practice one participates in. Being good as in being virtuous addresses the level of the character of individuals or communities. It is about living a good life.

Taking responsibility means trying to do good with respect to all four meanings of good. Consequently, accepting a forward-looking responsibility implies that one is willing to do more than simply fulfill the tasks or jobs according to the established moral rules. Taking a forward-looking responsibility means using your capacity of practical reasoning to enquire what a desirable cause of action is regarding these four goods. Thus, the argument "I am just doing my job" or "I am just fulfilling my tasks like I am ordered to do" is not a sufficient response to the question of why it is desirable to do something. Though it is important that practices and professions come with established moral rules, many authors on ethics and technology have recently argued that engineers and scientists should not adopt their given tasks passively, but adopt a proactive, forward-looking attitude (for example, Bird 1998; Lenk 1993; Mitcham 2003; Pritchard 2001; Richardson 1999; Ropohl 1993). These authors tend to focus on the responsibilities of engineers and scientists, but unfortunately they do not focus on responsibility for the social role of technologies.

It is also not self-evident to argue that taking responsibility for technologies implies being proactive using the capacities of practical reasoning and moral imagination. Rule-based understandings of responsibility are far more common, especially in daily life (Applbaum 1999). An often-used argument in favor of a narrow interpretation of responsibility that focuses on acting according to the established rules of the tasks and professions is the need for compartmentalizing society (Applbaum 1999). To uphold the high standards of working and living in complex social structures, people fulfill highly specialized tasks. According to this argument, professional compartmentalization implies that we also need moral compartmentalization, as some tasks require special moral norms. For example, everyone has the moral right to have a lawyer defend him or her against criminal charges. For this, lawyers are allowed to plead not guilty while they are aware that their clients committed the crime. Thus, people accept that lawyers lie while they generally see lying about such substantial issues as immoral.

The problem is, Applbaum explains, the widespread confusion between the descriptive elements of roles and the prescriptive elements of roles people fulfill (Applbaum 1999, p. 58): "a role simply is what it is, not what it ought to be." Therefore, deducing uncritically moral norms from the roles (or tasks and professions) is undesirable. One should not limit oneself to obedience to the formal and informal rules of the role. As Applbaum explains, being good at a certain task or profession does not mean that the task or profession itself is desirable. For example, one can be a "good" executioner in the sense that one is able to execute convicts efficiently while giving a great show to the audience. But, it is not said that it is good to be an executioner (Applbaum 1999). The same argument can be made for instance about nuclear scientists. One can be a brilliant physicist who builds a superb atomic bomb, but that does not answer the question of whether it is good to be a developer of atomic bombs.

Applbaum's argument is closely connected to the different meanings of good MacIntyre distinguishes. We can be bodily fit, and good at performing certain tasks, but that does not mean that we also fulfill a socially desirable role, nor does it imply that we are acting virtuously. MacIntyre promotes the argument that people reflect on all four meanings of being and doing good (MacIntyre 1999). He contends that the rules of a certain practice arise traditionally from shared social convictions, but that does not imply that these rules are good. Here, his practice oriented ethics seem to overlap with Applbaum's analysis on responsibility questions in relation to professions. Applbaum argues that professions should understand themselves as (Applbaum 1999, p. 60): "a calling that could have been otherwise, but argue why it still is a calling worthy of being answered by a reflective practitioner."

It is not enough to focus on established professional rules because we want to get a better grip on how new technologies will mediate our future lives. There are no suitable rules for imagining and evaluating the mediating role of new technologies. Of course, the social importance of moral codes, contracts, laws and the like is undeniable, but established rules can ask too little, too much, and are not suitable for dealing with the future social role of new technologies.

For a good action, following the established moral rules can be insufficient. Taking responsibility also implies enquiring whether following specific orders, guidelines or customs is desirable (MacIntyre 2007). In practice, a minimal understanding of professional responsibility that focuses on following the rules may lead to responsibility avoiding arguments such as "I acted according to the rules, so I did nothing wrong" or "everyone was involved in this". But, the fact that an action is not forbidden, does not imply that the action is good. Also, telling people that they should follow the rules can ask too much of them, especially when these rules conflict with their conscience. Think about the moral obligation of soldiers to obey orders. Such an obligation asks too much of them if they are told to kill enemies when they know the war is over.

But, perhaps, most importantly for our question is that falling back on established moral rules is difficult when developing new technologies or when employing technologies in new manners. In such situations, people rarely discover themselves in typical textbook examples in which the established rules easily apply. It is often unclear which rules should be addressed. To take responsibility in such dynamic contexts, enquiry about what is good to do is necessary. Take as an example the early days of genetic modification: in 1973, four researchers discovered the technique of genetic modification in vitro (Cohen et al. 1973), and they readily admitted that the hazards for the laboratory personnel, the general public, and the environment were largely unknown (Berg 2004). They recognized that the established rules for biomedical research were not suited for regulating the new techniques, and in 1974, their concerns culminated in a call for a voluntary moratorium on genetic modification. This moratorium provided time for estimating the risks and establishing guidelines (Berg et al. 1974). Most scientists acknowledged the moratorium until the end of 1976, when new guidelines were developed.

Where a forward-looking responsibility for the social role of new technologies is concerned, people need to be directed on what might happen in the future. A rule-based ethics is less suitable for imagining what might happen than is life ethics, since life-ethics reminds us of moral imagination or moral creativity as an essential virtue for doing good (Coeckelbergh 2006; Martin 2006). Moral reasoning—and thus taking responsibility—requires moral imagination on three levels (Johnson 1993):

- People need their moral imagination to see what is morally relevant in the specific situation
- Moral imagination is essential for empathy
- Moral imagination is needed to understand what the options and the likely results of the different actions are.

Especially for discussing new technologies or new employments of technologies, the virtue of moral imagination is important since within such fully new types of situations, it is not directly clear which details are morally relevant. Furthermore, the affected people will often remain "unseen", and the mediating role of the technology will be unknown (see also Chap. 6). Thus, being a responsible practitioner equals being a reflective practitioner who uses his or her capabilities of practical reasoning and moral imagination. But what does it mean to be a reflective practitioner?

4.4.2 Taking Responsibility and Practical Reasoning

As MacIntyre makes clear, to be a reflective practitioner you have to understand yourself as a moral agent with a distinct moral identity that does not equal your profession (MacIntyre 2006a). To understand yourself as a moral agent means to realize that your identity is distinct from the identity of others and that you are not merely your profession. To put it simply: you are not only an engineer, but you are also a civilian, someone's wife, a mother, a friend, a member of a chess club, and so on.

However, the relationship between the division of labor and the division of morality is not simple. The fact that someone has a certain role to play in society does not imply that he or she can just ignore moral responsibilities. Suppose a marketing officer finds out that within a certain price bracket teenagers are likely to choose a headphone that produces a louder sound while maintaining a reasonable quality. The task of the marketing officer is to make sure that the company makes as much profit as possible. Therefore, he or she should ask the engineers to design a headphone that produces an "earsplitting" sound. However, the same person may also be a parent responsible for the physical well-being of his or her children. Providing children with headphones that are likely to damage their hearing would be irresponsible. So, why does the marketing officer not have to take the responsibility to protect the hearing of other teenagers who are buying the headphones?

Important for a MacIntyrean philosophy is understanding that your identity does not equal your professional role, your cultural background or religion. Being responsible implies being able to have rational confidence in one's *own* judgment. Of course, judging is an activity deeply embedded in the techno-social networks. Reasons are provided by biological, social and technological factors, and to develop the capacities needed for practical reasoning one needs others. But this does not mean that we should just blindly obey others.

We should be careful not to make our-selves passive victims of our community (MacIntyre 2006a). In Carl Mitcham's words (Mitcham 2003): "this is role responsibility not as passive acceptance but as active agency, recognizing the extent to which we create roles at the same time that we are created by them". Moral imagination is implicitly stressed by the idea that moral roles are created by the people themselves. These authors teach that people do not simply have a social role: they co-create their own roles. This implies that before adopting a certain profession or task, one already has the responsibility to enquire whether the profession or task is part of a practice that is committed to the promotion human and environmental flourishing. Only in this way, can you make the actions your own, and take responsibility for them.

The final aim of MacIntyre's reflective analysis can be reformulated as an effort to empower people to take responsibility for their actions and lives. This also applies to people involved in techno-social change: for developing good technologies, all four meanings of good and the capacity of practical reasoning are essential. "Fundamental to MacIntyre's view is that one's status as an agent is bound up with one's capacities for practical reasoning" (Pinkard 2003, p. 182). As a consequence, this practical reasoning has to transcend the practices, although "the capacities cannot be understood outside the social and biological contexts in which they are realized". Taking responsibility as practitioner means using your capacity of practical reasoning to decide which particular occasion would do the most good with regard to the four different meanings of good.

Practical reasoning does not require the atomistic rationality of deontology and utalitarianism Latour argued so cogently against. There is no abstract "good" people can direct their actions to: there are only specific goods we aim to realize and to which we direct our practical reasoning. We do not start thinking as a tabula rasa: we have been molded by our culture, upbringing and experiences, and the moralities of the practices have taught us what to do. But we should not be too uncritical toward these established moralities either: being responsible means to be critical toward, and to rely on other people. We can only deal with our dependence by acknowledging it (MacIntyre 1999, p. 105):

> By independence I mean both the ability and the willingness to evaluate the reasons for action advanced to one by others, so that one makes oneself accountable for one's endorsements of the practical conclusions of others as well as for one's own conclusion. One cannot then be an independent practical reasoner without being able to give to others an intelligible account of one's reasoning. But this account need not be in any substantial sense theoretical.

Becoming a rationally less-dependent human being involves being held responsible by others for one's reasoning. People need others to develop the capacities of "independent" practical reasoning by evaluating, modifying or rejecting practical judgments. This is about asking whether the actions and attitudes that are considered to be good by the established morality, are really good actions or attitudes. For this, imagining realistically alternative futures is important. Here again, some resemblance with Latour becomes visible. Although Latour does not speak of acquiring virtues, he does explicitly state that ethics should be about making good choices, while acknowledging that all agency is embedded.

By focusing on our capacity for practical reasoning, MacIntyre works to show how we can take responsibility by making our choices really *our* choices (MacIntyre 1999, p. 111): "Without such virtues …, we will not only be deficient in discharging our responsibilities, but we will also be unable to deliberate adequately with others about the allocation of responsibilities." But MacIntyre does not study technological influences on human actions. Thus, the question remains, how does such an understanding of responsibility relate to taking responsibility for the social role of technologies? On one level, this means adopting a practice oriented idea of responsibility as described above and incorporating the undesired consequences or risks of the technologies. At the other level, however, the unexpected social role of the technologies need attention. How should we proceed?

4.5 An Ethics for Taking Responsibility for the Social Role of Technologies

The first challenge was to see whether we could develop an understanding of human moral agency that is based on aims, desires and intentions, without presuming free, atomist and rational human beings. In other words, can we be responsible human beings? Are we not too limited in exerting our free will? Can we know what is morally right to do? Several elements of Latour's and MacIntyre's views have been adapted to sketch a positive response to these questions.

First, there is the question of knowing what is right to do. Both Latour and MacIntyre claim that there are no objective standards for moral norms. As explained, Latour finds norms in the way people and technologies mediate each other's actions. MacIntyre explains that morality arises in traditions and practices. The two findings are complementary as long as we realize that being normative has a different meaning in Latour's and MacIntyre's vocabularies. Latour refers to the ability to direct action, while MacIntyre discusses the ability to have moral beliefs and virtues. Obviously, though both technologies and people have the ability to direct actions, only people can have moral beliefs and virtues. Technologies may be social factors, but only people are moral actors.

Second, even though Latour and MacIntyre agree that moral arguments may lose their efficacy in public debates because moral principles or rules are used in rhetorical power games, MacIntyre shows that on the level of practices moral arguments are influential. In practices, ethical arguments are not just emotivist remarks or means for a power game. Our morality is the result of more or less coherent worldviews that provide practical guidance for succeeding in the practices. Latour argues that moral principles are weak and that technologies are better at moralizing than ethicists. But MacIntyre shows that people need moral beliefs and virtues to flourish and to cooperate successfully. In the practices, he argues that people tend to strive for practical and moral "excellence". Though it is more realistic to argue that in the practices on average people strive for a sufficient quality level and not for excellence, still this argument provides an important motivation to be moral. We will not succeed in our communal aims—that is acquiring internal goods connected to the practices—if we do not share enough moral beliefs and virtues. Note that this is also true in taking responsibility for the social impact of technologies: if we fail to work out communal aims, moral beliefs and virtues, we will not succeed (see also Chaps. 6 and 8).

Third, since moral norms are not objective facts, but result from interactions in traditions, practices and networks, morality is pre-given, but not fixed. Networks are fluid and traditions and practices are also perceptible to moral change. On the one hand, this poses a problem for responsibility issues as it is less clear whether the norms and values of the practices can be taken as a moral guideline. On the other hand, flexibility is needed for taking responsibility as we need the freedom to reflect on what is desirable to do. It prevents people from arguing that "it is good since it has always been this way" (which is a natural fallacy).

Fourth, we are dependent creatures: MacIntyre recognizes our biological condition and social interdependence, but that does not mean we cannot take responsibility for our actions. MacIntyre points out that even though we are biologically vulnerable and socially interdependent, we still have the capacity of practical reasoning which enables us to take responsibility. Latour explains that we are dependent from other agents—people and technologies—that mediate our actions. Both views should be reconciled, arguing that we are biologically vulnerable, socially interdependent and technologically mediated. But can we still be responsible if we are dependent on so many factors?

MacIntyre empowers people to become more responsible by teaching how to make their actions their own, while acknowledging at the same time that our social environment and biological condition provide reasons for our actions. MacIntyre explains that there can only be a clear relationship between our actions and their consequences, we can only act more freely, willingly, and knowledgeably if we deliberate with others, using our capability of practical reasoning. In this, just like Latour, MacIntyre acknowledges that being part of a social network implies that others exert their power on you, but MacIntyre explains how we can work to take responsibility. However, he does so without referring to the social role of technologies, and so the inevitable question is: can we also take responsibility if technologies are acknowledged as social factors? This question resembles the second challenge, which can be reformulated as follows: can we reinterpret Latour's mediation theory in a manner that enables us to use our capacities of practical reasoning and moral imagination to steer the social role in a desirable direction?

To meet this challenge, Chap. 5 aims to reformulate mediation theory, and Chap. 6 presents an associated model of moral responsibility. These chapters are based on the following elements of Latour's and MacIntyre's teachings:

1. The social sphere can be partly described as a fluid network consisting of humans and non-humans that interact
2. Practices are part of these networks and provide virtues and moral beliefs needed for successful human interactions
3. Humans are biologically vulnerable, socially interdependent and technologically mediate, which gives them reasons for action
4. For being a moral responsible agent, humans need their capacities of practical reasoning and moral imagination

Chapter 5
Human Practices in Technological Contexts

After having dealt with the first challenge in the previous chapter, this chapter focuses on the second challenge: conceptualizing different forms of technological mediation to understand how technologies alter human action on a more detailed level. Meeting this challenge calls for reformulating Bruno Latour's mediation theory so as to incorporate insights from Alasdair MacIntyre as outlined in the previous chapter. Most important in this respect is MacIntyre's notion of practices, his idea of being biologically vulnerable and socially interdependent, and his distinction between practical reasoning and having reasons.

5.1 Reinterpreting Technological Mediation

The previous chapter argued for understanding practices in the context of actor networks. In techno-social networks, we perform coherent and complex forms of socially established cooperative human activity through which virtues are cultivated. Of course not all organized activities are practices, but in many activities we strive for doing good as in activities such as parenting, teaching, engineering, or painting.

MacIntyre's and Latour's views of human agency are future integrated by noting the extent to which people are biologically vulnerable, socially interdependent and technologically mediated. However, Latour defines agency in terms of the effects an entity (technology or person) brings about in the networks, while MacIntyre defines agency in terms of virtues. Consequently, Latour sees both technologies and people as agents or actants, but MacIntyre would not call technologies agents. MacIntyre makes, based on the recognition of virtue, a strict distinction between the action by things and humans (MacIntyre 2007, p. 201):

> The individual qua individual appears not only in managing the transitions from one role to another, but also, as I suggested earlier, in the role-playing itself. There are some roles that may seem purely mechanical, since the individual who plays the role always be replaced by a machine: where there was once a ticket-seller, there is now a ticket-machine. But the

K. Waelbers, *Doing Good with Technologies: Taking Responsibility for the Social Role of Emerging Technologies*, Philosophy of Engineering and Technology 4,
DOI 10.1007/978-94-007-1640-7_5, © Springer Science+Business Media B.V. 2011

> ticket-seller always faced choices that machines never confront: how to play her or his role, cheerfully or sullenly, carelessly or conscientiously, efficiently or inefficiently. And for all roles, the way in which the role is enacted presupposes not only an answer to a question posed to and by the role-player: "How is it best for me to play this role?," but also to such further questions: "By what standards am I to judge what is best?" and "Should I continue to play this role in this way?"

As a consequence, MacIntyre's understanding of agency does not apply to technologies. Artifacts simply lack virtues, moral beliefs, and the capacity to reason. But this still allows room for acknowledging that technologies play an important social role. Simply put, we can bridge the two authors by arguing that both technologies and people are *social factors*, while only humans are *moral actors*. With this reinterpretation, ANT poses a problem for MacIntyre's ethics since Latour argues that the social roles of technologies can change intentional actions into mere behavior (Callon and Latour 1992, p. 361):

> What is interesting, though, is that campus managers decided to shift the program of action "slow down cars on campus" from a culturally learned action to a mere piece of behavior—the physical shock of concrete bumps on the suspension of the cars. The program of action: "Slow down please for the sake of your fellow humans" has been translated into another one: "protect your own suspension for your own benefit." Are we not allowed to follow this translation through? Who made the move from action to behavior, from meaning to force, from culture to nature? We the analysts or they, the analyzed? Who or what is now enforcing the law, the standing or the sleeping policeman?

Latour argues that when speed bumps are built to slow down the traffic, slowing down is not a *moral* action in the sense that the action is not motivated by the desire to do good (Callon and Latour 1992). The original program of action "slow down please for the sake of your fellow humans" has been translated into "protect your car's suspensions". The outcome of these actions may be the same, and therefore Latour does not see the problem of ignoring differences between humans and technologies. But there can be a real problem when reasons for actions shift substantially. Technologies such as the speed bump can shift the starting point of our reasoning from caring for others into self interest.

One could argue: "So what? As long as people can safely cross the street, everything is fine, isn't it?" With speed bumps, this may indeed be a suitably pragmatic point of view. But often we also want people to be motivated by the desire to do good. For instance, some internet services offer to send yearly birthday and Christmas cards to loved ones. For forgetful people, this may seem to be a great solution. But what does it mean to receive a yearly card that was sent by some computer program? The virtuousness has disappeared in such cases.

So, the social role of technologies can be problematic in that we do not perform moral actions but show mere behavior: practices are no longer practices in the sense that they serve moral goods. And even though not all technologies alter human actions into mere behavior (many technologies may do just the opposite since they provide more options for actions, see also Sect. 5.3), a Latourian approach does not draw attention to our intentions: it describes intentional actions as if they were merely behavior that arises from human-technology associations. It is not the case that we must be completely free from external influences to be moral agents. Chapter 4 showed that MacIntyre's theory offers a valuable account of human agency in which our reasons for actions are biologically and socially given. We need food, water, fresh air, housing and clothing. We can suffer from diseases and mortality, and consequently, we are dependent on the goodwill of others to cooperate with us. We cannot survive, let alone flourish without others. But even though we are biologically vulnerable and socially interdependent, we are still moral beings in the sense that we can acquire the four goods by using our ability of practical reasoning. So, although biological conditions and social surroundings provide us with reasons for actions – and therefore limit our freedom – we can still reflect on these reasons, our actions and lives. Based on this reasoning, we can aim to alter course and become more responsible, even though we are not fully autonomous. Note that we are not always aware of our reasons: We do many things unreflectively and we would not say that we exert our agency when we do something unknowingly. An itching mosquito bite can cause unconscious scratching. But if one starts actually thinking and realizes that it is better not to scratch for it only makes it worse, not scratching becomes exerting agency, while the act of scratching was not.

Here, the comparison between being socially dependent and being affected by technologies can be made: other people can change your reasons for action, but as long as you are able to use your capacity of practical reasoning on those reasons, you can still take responsibility. This chapter argues that the same can be said for technological mediation. Technologies may alter our reasons for actions, but – as long as we are not discussing mind altering technologies such as psychopharmaceuticals and deep brain stimulation – they do not alter our capacity for practical reasoning.

A reason is a basis or motive for an action: it is why someone does something. Providing a reason is giving an argument, telling why one did or plans to do something. Reasons are those motivations for actions that can be reflected upon by people. MacIntyre argues that humans have a reason for action if he or she perceives that the action fulfils some aim. For instance, being thirsty is a reason for drinking some water. We do not have to be conscious of the motive for action, nor does a motivation for an action need to be explicated before it can be called a reason in the MacIntyrean sense of the word. But in order to take moral responsibility for one's actions, it is needed to become conscious about one's reasons, to explicate them and to reflect on them.

To be able to reflect on our reasons for actions (which is needed to make our actions entirely our own), we need to understand how our reasons are changed by technologies. Exploring how technologies mediate our actions by altering our reasons for action requires a revision of Latour's idea of mediation. Latour focuses on the outcomes of actions and pays little attention to reasons. His interest is on how our doings are mediated by technologies. ANT does not provide an understanding of who we are and why we do certain things. In other words, it does not provide any insight into our agency, and therefore it limits our understanding of how we can take moral responsibility.

His output oriented approach provides a suitable anthropological method to study socio-technological change from an external perspective. It clarifies how accumulative interactions between humans and technologies lead to new, unexpected outcomes, but it fails to make explicit how technologies mediate our actions and how we can regain control. By addressing the idea that technologies mediate our reasons for actions, which are evaluated in a process of practical reasoning, we can empower people to take moral responsibility.

Additionally, the idea of technologies and people "merging in action" in the network does not coincide with our experiences. It is true that people behave differently when they take their bikes for a small tour, than when they go for a walk. For instance, when walking, people from the Southern Dutch countryside greet everyone, often adding some friendly remarks about the weather. But, when cycling, they often only say hello to the people they know, and ignore strangers. Walking with at a pace of about 3.5 km an hour grants people plenty of time to salute everyone in a friendly manner. Biking at 20 km an hour, gives people far less time to greet others and make small talk. So indeed, the bike is mediating people's behavior. But, this does not mean that we are merged in action with the cyclists in the sense that the bicycle and the bike become a new agent by composition?

Our actions are not mediated by technologies in the sense that agency is the result of human and non-human associations. Instead, technologies mediate our actions by altering the reasons behind them. In the example of the bike, this simply means that you have less time to talk to other road users when you bike quickly. The reason behind the decision "to greet or not to greet" comes down to the matter of having time, and not of having become a new, merged agent with the bike. One could easily slow down a little and greet other people. Adding the distinction between having reasons and the capacity of practical reasoning (or pragmatic reflection) to mediation theory makes this theory more in line with everyday experience.

This chapter explains how technologies mediate our reasons for actions, and aims to explore what this means for our ability to take responsibility. This exploration is based on the distinction between three types of reasons for actions that together provide the input for further reflection on being the distinction between "is", "can", and "ought". The first type of reasons is formed by our factual beliefs. What we believe to be the case is closely related to how we see the world. The following section (Sect. 5.2) starts with analyzing how Peter-Paul Verbeek combines Latour's idea that technologies mediate our actions with the post-phenomenological notion of Don Ihde which explains that technologies mediate our perceptions of the surrounding reality.

Second, our reasons for actions are related to what we can do, or put differently: we also act in certain manners because we have options to do so. On the one hand, options for action limit our capabilities, but on the other hand, people often want to expand their options. Technologies mediate these kinds of reasons because they provide new, remove former, or alter existing options. Section 5.3 explains how technologies mediate our options for action.

Third, our reasons are based on our moral beliefs regarding moral values, norms, and virtues. Tsjalling Swierstra and Arie Rip have explained how new and emerging technologies such as nanotechnologies change our moral beliefs (Swierstra and Rip 2007). Section 5.4 offers an account of how technologies mediate our morality.

Finally, the last section (Sect. 5.5) analyses what this all means for the possibility of taking responsibility for the social role of technologies.

5.2 What Do We Perceive? Technological Mediation of Factual Beliefs

Verbeek 's *What things do* (2005b) was an important contribution to the debate on the social role of technologies. Verbeek distinguished analytically between the existential dimension and the hermeneutical dimension of mediation. The first form refers to the mediation of action and the second refers to the mediation of perceptions. Verbeek argues that Latour's view of the "existential dimension" of technological mediation does not describe the social role of technologies to the fullest extent. Therefore, Verbeek has elaborated on Latour's theory, and added with the help of Ihde's post-phenomenology, the hermeneutical dimension.

Verbeek's view is more comprehensive than either Latour's or Ihde's approach. He presents these two forms of mediation as an analytical distinction between two parts of the same process: it is not two processes that may play a role, but two sides of the same coin. However, this section argues that the hermeneutical dimension provides a deeper understanding of mediation of action since it explains how certain reasons for our action change. So it is not two sides of the same coin, but causal mechanisms: when our perceptions of surrounding reality changes, we have reasons to act differently. But the other way around this does not hold true: if we act differently, we rarely get different perceptions of reality.

Just like ANT, Ihde's post-phenomenology is an influential approach that aims to understand the broad social impacts of technologies by overcoming the subject-object dichotomy. But regardless of their similarity, post-phenomenology is often understood to be opposed to ANT (Ihde and Selinger 2003; Latour 1991). Still Verbeek has worked to integrate both approaches in an overarching post-phenomenological account since he argues that their views bridge the subject-object dichotomy in similar ways. Phenomenology overcomes the dichotomy by explaining that reality arises through relations between humans and their surroundings: humans must be understood as constantly experiencing the (phenomena of the) world while realizing their own existence in this experienced world. Post-phenomenology adds to the view that these relations are mediated by technological artifacts. This hermeneutical dimension of mediation refers to the transformation of human interpretations of reality. According to Verbeek, reality arises in the technological mediation between humans and the world: "Technology mediates our behaviour and our perception, and thereby actively shapes subjectivity and objectivity: the way in which we are present in the world and the world is present to us" (Verbeek 2005b, p. 203).

According to Verbeek, the mediating role of artifacts should not be understood as intermediating between humans and the world. Instead, mediation constitutes both subject and object at once (Verbeek 2005b, p. 130): "Humans and the world they experience are the products of technological mediation, and not just the poles between which the

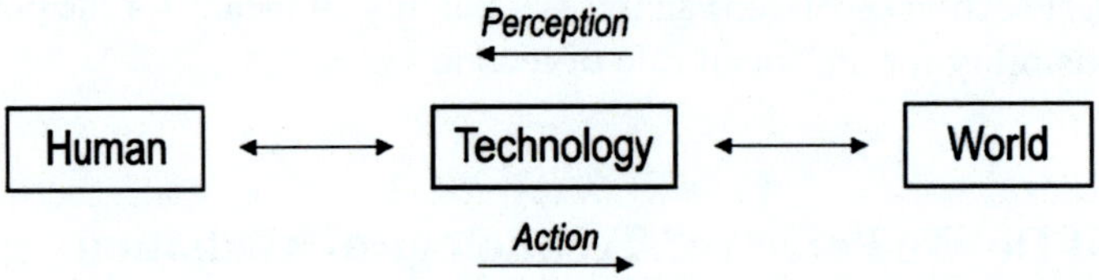

Fig. 5.1 Technological mediation according to Peter-Paul Verbeek (Source: Peter-Paul Verbeek, Lectures Industrial Design. Adapted from Ihde)

mediation plays itself out". Phenomenology focuses on explaining how we perceive or experience the world. Technologies mediate this experience. Post-phenomenology explains that experiences and meanings in our technological culture are regularly mediated. This idea is schematically presented in Fig. 5.1. The concrete examples Verbeek and Ihde mention include glasses, microscopes, and wheelchairs. They show that all these technologies transform the user's interpretations of the world.

Post-phenomenology teaches that we observe the world through technologies that transform our observations (microperception). A microscope makes it possible to see small particles or cells, but simultaneously excludes the larger context in which these materials were embedded earlier. Additional to the mediation of such microperceptions, Ihde (Idhe 1993) identifies macroperceptions. Macroperceptions consist of our worldviews, our understanding of the world. The invention of the letterpress, photography and film has tremendously altered the world-view of the average person. Of course these macroperceptions are informed by microperceptions, and vice versa. In other words: macroperceptions are interpretations of microperceptions. Since microperceptions are mediated by technologies, macroperceptions are also technologically co-shaped.

Verbeek explains that Latour's approach of mediation of action can be made commensurable with Ihde's view of mediation of experience. Latour, however, argues that Ihde's post-phenomenology is incompatible with his position, because it consolidates and reaffirms rather than overcomes the subject-object dichotomy. For Latour, it is essential to begin any human-technology interaction research by excluding the presumption of an a priori distinction between human and technological agents. Post-phenomenology, by contrast, accepts this distinction from the first moment onward, but explains that they come into being simultaneously and interdependently. Post-phenomenology does indeed take the two poles as a starting point, but this does not necessarily conflicting with ANT (Verbeek 2005b, p. 166):

> Post-phenomenology and actor-network theory want to do away with the gap between subject and object, but their different perspectives ensure that they do so in different ways... Phenomenology and post-phenomenology bridge the gap rather than denying it, by bringing to light the mutual engagements that constitute subject and object. Their perspectives are focused on the relation between humans and their world and, contra Latour, do not look 'from an externalist perspective' to describe how configurations of humans and nonhumans are continually arising everywhere. And from their perspectives it is indeed meaningful to make a distinction between someone who experiences and something that is experienced, someone who acts and a world in which action takes place – regardless of how interwoven and mutually constituted they are.

Explained in this way, Verbeek combined ANT and post-phenomenology as striving explicitly to develop a theory that distinguishes between artifacts and humans without adopting the subject-object dichotomy. He does this by recognizing the importance of the hermeneutic dimension: humans have perceptions with meaning, artifacts do not.

It is also possible to defend a less phenomenological interpretation of technological mediation, insofar as reality does not arise in the technological mediation between humans and the world. To some degree actions are mediated because our reasons for actions are altered. Put differently, the mediation of perceptions provides a deeper understanding of the mediation of actions. Instead of focusing on how mediation constitutes both subjects and objects at once in realizing themselves in the world, mediation of perception should be explained as the technological alternation of the interpretations that form reasons for people's actions. There is not necessarily reciprocity in the co-constitution of subject and object. As explained in the previous chapters, the primary agenda here is not to overcome the subject-object dichotomy: the main aim is to show how people as moral actors can take responsibility for techno-social factors. For this, we should accept Latour's argument that subjects and objects are not what many ethical studies presume (autonomous rational agents versus neutral tools). But, we do not have to treat subjects and objects symmetrically to the full extent nor do we have to conclude that they mutually co-constitute one another in the strict post-phenomenological sense.

I argue that the mediation of our perceptions (or factual beliefs) should not be seen as an addition to the mediation of our actions: the change of factual beliefs explains why people act in different ways. In other words, the mediation of our factual beliefs provides reasons for our actions on which we can reflect. Factual beliefs are based on technologically mediated perceptions of the world. These perceptions can be direct ("I have seen it myself") or indirect ("the micro-biologist told me ..."). Regardless of whether the perceptions are direct or indirect, they are often technologically mediated and so technologies influence those factual beliefs that are part of our reasons for actions.

It goes beyond the scope of this study to discuss in detail the drawbacks of a phenomenological ontology for ethics. But post-phenomenology is limited in only recognizing perceptions based on our experience of phenomena. This restriction results from the argument that subjects and objects are co-constituted simultaneously, which has an important disadvantage.

It is hard to escape one's own perspective in a phenomenological worldview, and consequently, practical ethical reasoning (MacIntyre 2009a) and moral imagination become problematic. If reality – including moral reality – arises in the mutual co-constitution of people and technologies, how is it then possible to take some distance to reflect on the desirability of a situation? How can we be morally imaginative about the social roles of a technology, if our moral reality arises also from the interaction with that technology? Of course, we are not atomistic autonomous beings who can reflect rationally and objectively on the desirability of certain developments. However, as explained in the previous chapter, it is important to work to become more independent and to use one's capacity of practical reasoning and moral imagination. To it put differently, if objects and subjects are mutually co-constituted, how can we be morally creative? How can we use our capacity of moral imagination in post-phenomenology if theory does not offer many clues on how to adopt an ethical stance?

Post-phenomenology stresses that action and perception are closely linked, but does not pay much attention to the fact that we hold many views that are not co-constituted by action. For example, the idea that we should treat equals equally is not informed by experience. Although it is hard to perform actions that are not based on conscious or unconscious perceptions, actions based on perceptions are a common occurrence. The ideas of logic (A and –A cannot be true at the same time), indirect knowledge (things told by others), and transcendentally formed ideas (metaphysical hypotheses and beliefs about god or gods) are examples of views that inform our actions, but are not formed by them. In other words, the kinds of reasons for action are not experience based. Perception (being a kind of reason or ground) is always primary to action, while many actions do not alter our perceptions at all. This is an important distinction, for two reasons. First, it enables us to address other reasons for actions that may be mediated by technologies (see Sects. 5.3 and 5.4). Second, it creates room for practical reasoning about these reasons, which is essential for taking responsibility (see Sect. 5.5).

Thus, in short, adopting a strict idea of simultaneous co-constitution of perception and action (of object and subject) is where post-phenomenology is deficient. It may be the case that in certain situations "reality arises within the technological mediation between humans and the world". But this is not always the case: Internet could serve as a great example of a technology which mediates shaping and perceiving (virtual) reality simultaneously. Here, Latour's view that for studying techno-social networks a certain kind of realism is required, conflicts with phenomenology.

But regardless of whether one agrees with Verbeek's or the previous arguments, we may conclude that a change in the interpretation of reality can provide new or other reasons for action. Sometimes we will be unconscious about this change (consider for instance the many signs to which we respond automatically), but conscious alterations also take place. A well-studied example is the effect of the microscope on our perceptions of hygiene and diseases (Barnes 2006). When microscopes allowed us to see micro-organisms (microperception), people started to acknowledge the possibility that these tiny "beasts" could be the source of infections. They realized that the solution may lie in avoiding contact with these creatures (macroperception). This understanding changed the view of people on what could be considered safe or healthy and what could not, and consequently they changed their actions regarding hygiene and sanitation. It is not hard to see that the observation of microbes altered the interpretation of causes of diseases, and therefore changed the reasons of people's actions.

Finally, it is also useful to point out that it is not only our perceptions that inform actions: commonly, we have other reasons as well. Technologies do not only mediate perceptions, they also mediate options for actions and moral beliefs (our evaluation of values, norms and virtues). The next two sections explore these two other types of reasons for actions that are technologically mediated.

5.3 What Can We Do? Technological Mediation of Options for Action

Determinists often argue that new technologies restrict our options for actions. Although Latour is definitely not a determinist, many of his examples show that the options for actions are reduced by technologies. A heavy weight on a hotel key makes it less convenient or even impossible to put it in your pocket. A sleeping policeman makes the option of speeding considerably less attractive. An automatic door groom makes it harder to pass through the door with a wheelchair. And perhaps that is why many authors have taken Latour's work as a source of inspiration to plead for deliberately designing moralizing technologies.

Early philosophy of technology often stressed that traditional crafts were highly appreciated in society and were thus rewarding to execute, and that technologies transformed these trades into uninteresting activities performed by unskilled labor. In 1811, the Luddite movement was the first to argue that valuable social tasks disappeared because of new technologies. Their first argument was that the Industrial Revolution threatened their subsistence, but their criticism also had a more fundamental side: they resisted the idea that all labor would become unskilled and mechanical (Jones 2006) and feared that workers would be reduced to slaves of machines. A similar point is made in the 1930s movie *Modern Times,* in which Charlie Chaplin suffers a nervous breakdown from working on an assembly line.

Additional to pointing out mass production and assembly lines in the twentieth century, social pressure became recognized as a reason why new technologies could lead to a limitation of our options in daily life. If everyone is using mobile phones and e-mail to communicate, you cannot reject these technologies, without becoming an outsider and, most probably, unemployed. Furthermore, a new technology can compete and even diminish former options, and therefore some authors argued for a technological evolution theory, for example (Basalla 1988). New technologies replace technologies that are regarded less efficient, less comfortable or less functional, and only the "fittest" technologies survive, according to this view. The case of the zeppelin is a well-known illustration.

Finally, technologies alter our physical adaptation. Edward Tenner analyzes for instance how shoes decrease our options for action by mediating our physical needs (Tenner 2004, p. 53): "People in industrial societies may assume that the tender sole would soon be ravaged by its environment without the protection of footwear. In fact, wearing shoes creates this sensibility." Toward the end of the nineteenth century, it was common for example, for many Irish children living in the countryside to walk to school barefoot. When walking barefoot for a week, protective thickened interstitial tissue forms and the skin becomes horny, protecting the foot from injuries and cold. Wearing shoes every time you step outside makes the tissue perhaps better looking, but also softer and creates the need for protective footwear. In other words, the option for actions become limited: we are no longer able to walk barefoot for many miles if we are used to wearing shoes.

It cannot be denied that options for action disappear because new technologies. Traditional philosophy of technology has provided valuable insights in this regard.

But today, many philosophers of technology understand technologies as means that also increase options for action. Many new technologies – such as wheelchairs, computers and internet – promote new opportunities for people who are less able to participate in society. Due to the mediation of technology, they become more mobile and achieve more means to communicate with other people and as a result alter their views on the good life. For instance, the virtual world *Second Life* also gives disabled people a chance to appear and communicate just like everyone else, overcoming the difficulties of real-life. Photographer Robbie Cooper has created a book in which photographs of real people and their avatars on *Second Life* were published next to each other, accompanied by short interviews (Cooper and Spaight 2007). The book showed that a substantial group had social or physical challenges and used the medium to present themselves in a way that they saw as being more representative of their inner selves. As Cooper and Spaight's research revealed, most players were seduced by the endless possibilities to improve their appearance on the net (see also Chan 2007).

Along with decreasing or increasing our options for actions, technologies also *alter* them. For instance, about a hundred years ago, people had only one way to write a letter to friends or family members living abroad: writing on paper. Today, we often use e-mail to stay in contact, and this alters actions. If the only option you have is to write a letter on paper that will take a long time to arrive and would probably be kept for a long time, you will sit down, reflect and try to do your best to write a beautiful text. E-mail however, is more instant and has the option to respond immediately. We also get so many emails that we delete most of them after some time. So, we tend to write short, quick answers, instead of elaborate letters. If a quick answer is not enough, it is at least polite to let someone know that the message has been received and that a longer answer will follow. The technological option to respond directly is available, and so we are more or less bound to respond quickly. Responding to an e-mail from family or friends is often no burden, but all these new forms of communication are not only used by friends. There are also other settings where the idea that you can respond immediately implies that you ought to act immediately.

So, by changing our options (increasing, decreasing and altering them), technologies mediate our reasons for action: if our options change, our duties can change with them. Often, technologies increase options for action, which encourages people to assign more duties to each other. If there is no medicine or vaccine for a disease such as leprosy, the rich cannot be asked to help cure people in poorer regions. But, with contemporary technologies, it is easy and affordable to fight disease, and the question can be posed as to whether or not it is the duty of the rich to eradicate leprosy.

5.4 What Do We Want? Technological Mediation of Moral Beliefs

Latour extensively explains how human-technology interactions alter the world, but he pays no attention to how technologies alter the way we want the world to be. In other words, he does not explicate the technological mediation of our moral beliefs (principles, values and virtues). Additional to the perceptions that form our factual beliefs, we have multiple beliefs about how the world ought to be. Moral beliefs are convictions about what is good to be and good to do in relation to the flourishing of oneself, other humans and sensitive beings in general. Based on these moral beliefs, we evaluate options and perceptions. It is often claimed that technologies can have a moral impact in the sense that they influence the quality of life, and the physical and mental well-being of people and other living beings, but theories that explain how technologies alter our understanding of the good life are rare.

How can we understand moral change due to technological developments? One of the rare accounts that addresses the technological mediation of morality is to be found in Swierstra and Rip's article on NEST-ethics (Ethics of New and Emerging Science and Technology). NEST-ethics is inspired by ANT, but offers something extra: it clearly identifies that because of technological change our morality changes, and vice versa (Swierstra and Rip 2007). NEST-ethics was developed for a better design of future scenarios for technology assessments. It rejects the commonly adopted premises of scenario builders that technologies change while morality remains fixed.

Without falling into moral nihilism, NEST-ethics aims to explore which facets of morality are universal and which facets are likely to change over time. Although this is an important step toward understanding the impact of technologies on the moral sphere, Swierstra and Rip explicitly recognize that the next step is to develop a theory of techno-moral change (Swierstra and Rip 2007), p. 19: "An important point, which remained implicit in our discussion of NEST-ethics, is the co-evolution of ethics and new technologies: while there are recurrent patterns of moral argumentation, there is also learning, shifts in repertoires, new issues coming up". During the Expert Seminar *Anticipating the Interaction of Technology and Morality* (2008) Swierstra again signalized that "we lack a theory of techno-moral change."

Here, this need is translated through the question of how technologies can mediate moral beliefs. MacIntyre's notion of practices (1985) can be used as a starting point to show how moral beliefs are technologically mediated. MacIntyre himself makes a connection between practices and technologies when he acknowledges that science and engineering are practices. It is not hard to see why Macintyre makes this claim: Science and engineering are complex, more or less coherent and socially well established multi-actor activities that require special technological skills, values and principles. Furthermore, scientists and engineers are said to strive for moral "excellence". In this pursuit also the production of moral norms is important, as Merton has started to explore (Merton 1973). Merton has identified four norms that scientists have to strive for in their work, knowingly: communality, universality, disinterestedness, organized scepticism. So, the idea that science and engineering are practices in themselves implies that their work is perceptible to moral reflections.

Note that here, the definition of practices is used quite narrowly. In daily language and in many philosophical and sociological writings, the concept is used in a much broader sense, referring to commonly occurring or institutionalized forms of human cooperation. Here, the concept is used in a MacIntyrean sence, though it is not presumed that practices always lead the production of internal goods such as art or music. MacIntyre's concept of practices can help to make clear what technologies do in relation to the moral sphere. It is important to recall that practices are not opposed to actions: instead, they are a special form of actions, knowing, cooperative actions that are complex but coherent, and in which people need the virtues. Technologies mediate practices, meaning that they alter what we believe to be virtuous and morally right.

There is also another relation between these practices and technologies, which becomes clear when understanding practices as parts of techno-social networks and thus co-shaped by technologies. In discussing practices, MacIntyre focuses exclusively on *humans*, but it is not hard to imagine the mediation of *technologies* in relation to practices. Socially established practices that are low tech are rare, though not imaginary (weddings are a widespread example), but most practices are mediated by technologies, to a larger or lesser degree. Creating art and teaching are just some everyday examples of practices in which technology plays an important role. Their nature has been reestablished due to new technologies. Art societies for instance are often first and foremost defined by their means, such as Montevideo (the Dutch Institute for Media Art), the US Enamellist Society, the Dutch Studio for Electro-Instrumental Music (STEIM), the Ceramic Arts Association of Western Australia and the British Computer Art Society.

Practices are not only organized around technologies: technologies can also mediate existing practices. Consider for instance the role of new technologies in the religious practices with the cell phone Ilkone i800 that is designed to support a Muslim lifestyle. Ilkone is derived from an Arab word meaning "universe". The hardware and software of the Ilkone i800 stimulates the user to pray. The phone has a digital compass that points to Mecca and the phone's alarm rings five times a day when it is time to pray. In this example the Ilkone i800 is far from passive, since it stimulates the user to devout Muslim actions. In other words, the device is mediating religious practice. Furthermore, after the alarm goes off, the phone automatically goes into silent mode for 40 min. This prevents the user and other Muslims from being disturbed during their prayers. Now, one can argue that the Ilkone i800 is designed to stimulate the user's practice of prayer, but as we have seen before, devices that are not designed to fulfill a social role also mediate human actions.

Practices can be technologically mediated in different ways. Existing practices disappear and new practices arise. Innovations can sponsor new practices, for example, the practice of architecture could only arise when a broader range of building technologies was available. Additionally, existing practices, like for instance professional wet-nursing, can disappear when key technologies are replaced by new ones, like artificial baby milk. These mediations are often rather eye-catching and have more to do with the change in options for action than with the change in moral beliefs.

However, practices can also be mediated on a more fundamental level. It is important to realize that practices are not only about what people do, but also about what they believe to be valuable, right or good. It is a mistake that agency is mainly about action, and Latour did not recognize this when he adopted his understanding of agency. He formulated an output oriented definition of agency on the results of action, on what acts bring about. But, human agency is not primarily about what someone does. Instead, it reminds us that actions flow from our reasons. We act in one way instead of the other because of who we are and what we think. In the practices internal goods are shared (MacIntyre 1985), which implies that by interacting, people develop at least some common moral beliefs. In other words, the values we hold are closely related to the things we do together.

As mentioned earlier, moral beliefs consists of ideas about acting good, which are the values, norms and virtues. Our moral values – for example the value of autonomy, privacy, or trust—are about what we believe to be important for human and environmental flourishing. Moral norms, or rules, explain what to do or what not to do to pursue those values: the Ten Commandments are examples of moral rules. Virtues are about the attitudes or character traits we need to adopt for having and living a good life (or to do the morally desirable thing). Being responsible or trustworthy are examples of virtues.

5.4.1 Values

With the hermeneutical dimension, Verbeek mainly focused on how we interpret the surrounding reality. Of course, this also has consequences for moral actions (Verbeek 2008b), p. 13: "After all, the post-phenomenological approach makes it possible to investigate how technologies help to shape human perceptions and interpretations of reality on the basis of which moral decisions are made." Verbeek explains how ultrasound mediates the relations between fetus and future parents: "it constitutes both in specific ways and therefore it plays a crucial role in moral decision-making."

Verbeek makes an important contribution to the debate: he shows the relationship between our morality and the technological mediation of our perceptions. But morality is not only a matter of how we perceive others. Perceiving others in a different manner (such as observing the fetus with the help of ultrasound) can lead to a change in our reasons for actions (we see how the fetus is moving and developing), but it can also change our moral values. Verbeek's description about perceiving the fetus differently is about our factual beliefs and not about values. In other words, he addresses how ultrasound changes our views on the question: "how 'human-like' is a young fetus?" But he does not go into much detail about how technologies mediate what we believe to be morally valuable (for instance, he does not discuss the question of whether technologies altered our answer to the question "is human life precious?").

Consider two examples of how new technologies (or new applications of technologies) can bring about a shift in moral values. The first example is the change in the values of privacy and security in relation to the cameras placed in public spaces. When in the early 1990s some Dutch city councils decided to install surveillance cameras at train stations, squares and on shopping streets, many people protested that this was a violation of their privacy. Nowadays – only 20 years later – people are complaining that there are not enough cameras in public spaces to ensure their security, while safety on Dutch streets has not substantially decreased and crime did not increase. People have become used to cameras observing them, and they have gradually come to value the idea that cameras are present to establish safety. Consequently, they have begun to value their privacy less, while valuing safety issues more. Moreover, since early 2010 all passengers boarding at Schiphol and Heathrow on transatlantic flights have submitted themselves to security cameras that can see through clothes, and even this has led to merely a small protest.

In this first example the balance between two opposing values shifts: privacy has become less important in favor of safety. But technologies can also alter the meaning of a value. For instance, the meaning of friendship is altering because of the internet. Thousands of new friendships that are started every day on Hyves and Facebook can hardly be argued as being genuine. Of course, occasionally such a friendship can grow into an enduring closeness, but most of them will quickly subside. If you are bored by someone or if a so-called friend becomes a bit irritating, you can just dismiss him or her with one click of the mouse. There is no stimulus for persevering with this kind of friendships. This is not to say that computer mediated real friendship is not possible at all. Some people meet their partner via the Internet and several authors have pointed out the added value of the Internet for finding and maintaining friendships (see also Briggle 2008). But when acquiring 35 new friends in 1 day, one can say that most of them are not real friendships and few will be enduring, even though personal information is often shared amongst Internet friends.

5.4.2 Norms

Beside values, norms (moral standards for actions) are also mediated by technologies. Technological artifacts are part of practices in the sense that they form, and are formed by, the moral standards of the practice. Innovations can either "sustain" or "transform" the moral standards of practices. An example of "sustaining" is the engineer who develops a hammer drill with the idea that using a hammer drill is typically a male job. The engineer designs a heavy machine with a large handle on the drill, so that it will fit into the large hand of the average construction worker, instead of choosing an alternative that is suitable for smaller hands as well. At that moment, the designer's belief becomes a self-fulfilling prophesy, since the drill will be unsuitable for the average female hand. This way, the drill has become a mediating tool: it "says" drilling can best be done by males and our view on how to divide the tasks is reaffirmed. If we would like to change this, the drill's design should be modified. Comparable examples are the special feminine mobile phones, cars and movies that are increasingly introduced onto the market. These technologies simultaneously recognize and reinforce gender differences.

Frequently, technological developments (for example information technology and television) are associated with the – unintended – global standardization of habits and viewpoints (Lash and Lury 2007; Levitt 1983). Those viewpoints claim that technologies affect our life world deeply, but the claim is rarely made that technologies alter our moral beliefs.

5.4.3 Virtues

Some philosophers of technology who have argued that our virtues are changed by technological developments, usually understand technology as an all-encompassing system. For instance, Marcuse (Marcuse 1964) has argued that in capitalist societies, technological development creates false desires and people become entangled in the network of production and consumption. According to Marcuse, this results in "one-dimensional" people who lack critical thought. Since we all adapt the same roles of producers and consumers, our lives and characters are becoming superficial. It is about what we have, and no longer about who we are.

Being one-dimensional is also explained as being "morally unimaginative". Günther Anders (Anders 1980/1956) described how technologies may affect our empathy in a macabre way when he discussed the bombing of Hiroshima. To drop a bomb, the pilot only had to press a button and he had to face neither the victims nor the consequences of the bombing. Without hearing or seeing the impact, he was able to kill millions of people, while, as Anders claims, listening to classical music. Due to technology, people become emotionally detached and social coherence declines, Anders argues. But his negative attitude toward technologies faced extensive criticism (Dijk 2000). Technologies such as television–which was often criticized by Anders–and the Internet have increased emotional engagement in many regional, national and international political debates on all kinds of issues. It makes the suffering of people in Third World countries visible and so stimulates empathy.

Pessimistic views like Marcuse's and Anders' do not do full justice to technologies: that is, altering what we believe to be virtuous instead of diminishing virtues. Foucault's description of discipline in schools makes clear that the classroom design and the chairs and desks enforce a certain physical posture for education (Foucault 1975), which encourages a moral posture or attitude. The bodily position is closely linked to the attitude required for learning. Only a few decennia ago, classrooms were designed in such a way that the students were forced to sit up and look at the lecturer or teacher. The rooms and furniture were not designed to stimulate communication, but listening. There was little room to move and only small desks were provided in order to make notes. This design was closely connected to what was then believed to be good education and students' attitudes were expected to be rather "one-dimensional".

Today, complex, multimedia rooms are developed for education in which large, wheeled tables and luxurious office chairs are placed. These surroundings are not only much more comfortable, but they are intended to stimulate a pro-active learning attitude. Students are no longer supposed to sit quietly and listen; they have to work on projects, engage in debates and communicate with others.

The virtuous present-day student is unique, pro-active, assertive, communicative and collaborative, instead of observational, timid, obedient, and solitary. This change in the appreciated virtues of study has been deliberately stimulated by the furnishing. Often, the mediation of our attitude and ideas of the good life have a more emergent character.

The example of the student classes illustrates how technological artifacts mediate our virtues, and it also shows that virtues are not interdependent on each other. Moreover, it points out that there is a close relation between our virtues and actions. The furniture determines the student's ability to adopt a bodily posture (a physical condition) which stimulates the attitude they occupy in relation to each other and to the teacher (which has to do with virtues). As a result, students are encouraged to act in a way that is considered virtuous in the educational practice. Teachers are also stimulated to change their ways. When the furniture of the classrooms is not suitable for cooperation projects (for instance because all of the benches are nailed in rows to the floor), teachers are less likely to provide students with cooperation assignments, for obvious organizational reasons. Also giving long lectures to students seated in a room designed for interactive projects is something most teachers would not choose as it becomes much harder to keep the students attention. Many elderly people often feel there is no coherence in contemporary classrooms, while students often remark that by working in groups their social coherence increases and they are trained to cooperate professionally.

Summarizing, we can say that new technologies, or new applications of existing technologies can alter our values, norms and virtues, or – to be short – our moral beliefs. The previous section discussed how technologies mediate two other types of reasons for actions: our perceptions and our options for action. If the argument is correct so far, what does that mean for the ability of taking responsibility for the mediating role of technologies?

5.5 Again: The Question of Responsibility

The second challenge formulated in Sect. 3.5 was to conceptualize the different forms of technological mediation, so we can come to understand how technologies alter human action on a more detailed level. The argument was that only if we understand technological mediation on a detailed level, can we work to take responsibility for it. If you want to influence the processes successfully, you need to understand the mechanisms of such processes, at least at a minimum level. For this aim, this chapter reformulates Bruno Latour's mediation theory and argues that our actions are technologically mediated because technologies alter the reasons that inform (consciously or unconsciously) our actions. Reasons for actions are commonly divided in three categories: is, can and ought, and technologies mediate all three of these types of reasons.

The mediation of the first category is well explained by Ihde and Verbeek. They argue that technologies mediate our observations and interpretations of the

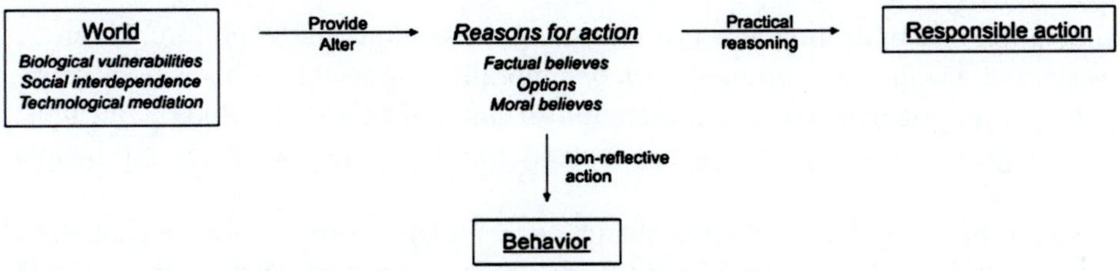

Fig. 5.2 Responsible action in relation to technological mediation

surrounding world. The second category – technologies change the options for action – is explained by addressing the commonplace idea that new technologies offer new possibilities. But as we have seen, technologies can also decrease the options we have or transform the existing options. The third category is introduced with Swierstra and Rip's Nest-ethics, and is explained by examples of how technologies change our moral beliefs.

Summarizing, technologies mediate our views of the factual, the possible, and the desirable. Of course, these three types of reasons are interrelated to each other: sometimes we ought to do something because we can, and often we only can do something because we understand some factual aspects. Furthermore, it is important to stress that we are often not aware of our reasons for actions. We act "a-responsible", and accepting and taking responsibility means aiming or trying to become aware of the reasons for our action, evaluate these reasons and alter our actions according to this evaluation.

For example, while I was writing the sentences above, I appeared to have drunk a large glass of diet coke. I remember the waiter putting the glass on the table, but I am not aware of the fact that I emptied the glass, nor that I was thirsty. So if technologies mediate our reasons for actions, it can simply be the case that we are not aware of it. The bodily position that has to be adopted on a chair is an example of how the chair manipulates your reasons for altering your pose, without demanding much awareness of the average user.

But although we are not aware of all our reasons for action, we can work to take responsibility for our actions by using the capacity of practical reasoning. I can be called a responsible agent, as I am able to reflect about my "diet-coke-drinking" and can decide that it is actually not wise to drink too much coke. The fact that I often tend not to reflect on these kinds of matters, does not mean that I am not able to take responsibility for these kinds of actions. In other words, I am acting non-responsibly when I do not reflect on my eating and drinking habits. Irresponsible behavior is for example drinking too much coke while realizing that it is bad for me. Taking responsibility implies using the capacity of practical reasoning to adjust your actions. Consequently, taking responsibility also implies that you can give a response to the question: why did you do that? See also Fig. 5.2.

Now, as long as technologies do not interfere with the capacities of practical reflection and moral imagination, technological mediation does not in principle, diminish our agency. Technologies such as psycho-pharmaceuticals, deep brain stimulation and brain implants can change our reasoning, and therefore, it can be questioned whether we can hold someone morally responsible if his actions originate from physical or chemical alteration of his brain (Verbeek 2009). But most technologies do not alter our mental capacity to reflect, and therefore, in principle we are still agents.

So far this study has worked to obtain a general answer to the second challenge. But does that entail that we have a full reply to the question of whether practitioners can take responsibility for the social role of technologies? To be able to take such a forward-looking responsibility, the last challenge still needs to be addressed. Chapter 3 explains that ANT is problematic for responsible action, because the techno-social networks constantly change, which makes is hard to imagine what the possible social role of a technology will be. We have to consider the question of whether we can be imaginative enough to foresee the possible outcomes of technological mediation. This does not require absolute knowledge about future developments (which would be absurd). To take responsibility, we should at least be able to make an educated guess about how new technologies or new application of technologies will mediate the reasons for actions of several involved practices. To do this we should be able and willing to use our moral imagination.

On the question of how to employ the virtue of being morally imaginative, we also stumble upon a second problem. How can we evaluate technological mediation if technologies mediate our moral beliefs? Swierstra and Rip's argument that when reflecting on future techno-social change, morality should not be considered to be a fixed factor (Swierstra and Rip 2007) was adopted for this study. Instead, morality is a variable, which changes when society and technology develops. It is, just like the rest of the network, fluid. As a consequence, one can wonder which norms to use for evaluating possible outcomes. The following chapter works to formulate some preliminary solutions to two problems regarding moral imagination.

Chapter 6
Tools for a Forward-Looking Responsibility

In this chapter, we arrive at the final question of this study: what would enable and support practitioners exercising a forward-looking responsibility for future social roles of technologies? Chapters 4 and 5 discussed the first two challenges formulated in Sect. 3.5. The first challenge was to show how people can accept a forward-looking responsibility for our actions while also rejecting the autonomous and atomist subject of the deontological and utilitarian theories. To take on this challenge, Chap. 4 explained how ANT can be morally enriched by recognizing that moral practices are part of the human-technology networks. Furthermore, it explores how people can make their actions their own in the networks and practices, even though these actions are biologically, socially and technologically embedded. The second challenge was to understand mediation theory in a manner that made it feasible to appreciate the different forms of causation. To meet this challenge, Chap. 5, reformulated mediation theory with Alasdair MacIntyre's work on practices, reasons and reasoning to show that technologies alter our actions by mediating the reasons for action—the perceptions, options for actions and moral beliefs.

The argument is that practitioners (individuals or institutes) can take responsibility by using their capacity of practical reasoning to reflect on these reasons. Thus remains the third and last challenge: to develop some practical tools that enable us to take a *forward-looking* responsibility. We need some tools to "open the black-box" prospectively to make informed choices. How can such tools be obtained? We cannot develop tools to predict the future in detail, and it is unlikely that we will find a "crystal ball" to inform us on the long-term consequences of our contemporary choices. Nevertheless, if we want to take a forward-looking responsibility—or, put differently, if we want to make well-balanced decisions—we should at least try to imagine what the possible social and moral impacts of a new technology or new use of technology can be. The results of such deliberations may not all become reality, but such exercises are important to explore desirable future human-technology interactions. Since we know from experience that human-technology interactions may have unintended and undesired outcomes, it is desirable to explore what might happen with a new technology or new use of technology. This way, we can

K. Waelbers, *Doing Good with Technologies: Taking Responsibility for the Social Role of Emerging Technologies*, Philosophy of Engineering and Technology 4,
DOI 10.1007/978-94-007-1640-7_6,

inform our contemporary choices. We have to make choices in current technological developments, and those choices are likely to have future consequences. So, even though we will never be sure, it is better to educate our choices and try to realize what we believe to be a desirable future, than to play dice hoping that things will work out fine in the end.

Thus, even though foreseeing the future is not an option, reflecting about the possible future mediation of technologies is worthwhile. We need to suggest ways how to do this in practice. How can we become more imaginative about the possible future social roles of technologies?

The question of becoming more imaginative about the possible future social roles of technologies can be broken down into several parts. First, it is important to distinguish the more descriptive aspects from the more normative or evaluative aspects. The more descriptive part of the enquiry can be decomposed into two parts. It would be helpful to have (1) a list of questions to structure the deliberation on the human-technology interactions, and (2) different kinds of research methods to explore these interactions. The possible social role of technology precedes the question of whether that possible social role is desirable. Thus, a toolbox for taking responsibility for the social role of technologies includes three types of reflection, which are discussed respectively in the upcoming sections. Section 6.1 introduces the first part that supports the structure for the enquiry by providing a step-by-step approach for addressing the interactions between the techno-social factors and moral actors. Section 6.2 presents the second part of the toolbox by explaining how the five questions raised in the first part can be discussed with the help of three different academic approaches. Section 6.3 discusses the third part of the enquiry by offering some preliminary means to evaluate the obtained insights.

6.1 Part 1 of the Toolbox: Questions to Address

This section presents the first part of the toolbox: the descriptive part which aims to structure the enquiry on the future social role of a technology (and thus to lessen blind spots or biases in the evaluation). For this, I developed a step-by-step approach that questions whether the social role of a technology commits to human and environmental flourishing. The rationale behind this approach is that practitioners often develop technologies for a reason that is allied to serving other practices. Superstores put foods in cool storage in order to keep it fresh for customers. Engineers are developing more efficient engines because there is a need for more economical and more sustainable automobiles. The engineers are embedded in the engineering practice, but their choices only make sense because of the use of the technologies in practices. The aim of engineering in general is to produce technologies that are useful in other social settings.

My starting point is that technologies should not only be useful to those other practices, but they should also support a good life (or flourishing) within these practices. Therefore, it is important to question (1) which other practices will be affected, and (2) what are the common reasons of actions within these practices. This information helps to see how these practices might be mediated.

Further the step-by-step approach should be directed at identifying possible interactions between the aims of the technology and the common reasons for actions of the involved practices. Therefore, when the aim of the technology is clear, questions are directed toward identifying the common reasons for actions of the involved practices. In other words, it asks for a description of the aim and functionality of a new technology or a new employment of technology and of the reasons for actions in the involved practices. Next, for discussing the future social role of technologies, the enquiry needs to point out how the involved practices may employ the new technologies and how their common reasons for actions might be mediated by such employment.

Here, a slightly different—or less high-minded—definition of practices than MacIntyre is used. MacIntyre's concept is limited by its focus on "noble" activities that aim to produce intrinsic goods such as art, music, and knowledge. But the development of morality is not limited to just these activities. We also have moral expectations of people in many of our activities, and for any good cooperation, virtues are needed. For example, the practice of doing business mainly produces external goods (money and power), but requires multiple virtues (such as being capable of trusting others and being reliable) and involves many moral beliefs (such as keeping promises and not breaking contracts). Therefore, I adopt a less demanding definition of practices: any common cooperative human activity that requires virtues and involves shared expectations is a practice.

From these starting points, the first (descriptive) part of the enquiry consists of a five-step approach:

1. What are the aims of the technology?
2. Which practices will be affected?
3. What are the common reasons for actions in those practices?
4. Given these reasons and given the existing technologies, what uses are likely?
5. How will these uses mediate the reasons for actions in the involved practices?

Ad 1. What are the aims of the technology? The first step is to explicate what the aims of the new technology or the new employment of the technology are. Technologies are made and used for a reason. Many technologies are meant to extend human power and ability (like hammers and airplanes), others aim at amusement or relaxation (like iPods and DVD players) and some aim to replace humans or human capacities (like care robots and calculators).

We should not only address the primary aims, which commonly refer to the functionality of the technology. Secondary aims are also important. For instance, the primary aim of shoes is to protect your feet. The secondary aim of shoes is often related to fashion. For many people, this second aim is at least as important as the first aim in making the decision of which pair of shoes to buy. Since technologies are made and used for a reason, one can morally evaluate these aims: Does the aim support human and environmental flourishing? Or, to put it a little less demanding: does the aim refrain from conflicting with human and environmental flourishing?

Ad 2. Which practices will be affected? Step one evaluated the aims of the technologies, but even when the intended aim of technologies is to support human and environmental flourishing, it can still conflict accidentally with that flourishing: unintended social roles are common. To enquire whether the technology will interfere with human and environmental flourishing, it is important to know which practices will be affected by the technology.

Mediation is not a linear process between just two entities: many actors are involved in many different manners. Due to the mediation in the techno-social networks, the initial action and the eventual outcome of the action may become dissociated in time and in physical and social space (Waelbers 2009b). In other words, when people act through a technology, it is not uncommon for their action to have an effect on practices they did not expect to be involved. Thus, technological mediation becomes hard to imagine. To deal with this problem, the second step consists of mapping out the involved practices. This entails the current practices as well as those of the foreseeable future.

In the future steps, practices should not be considered to be too uniform: different kinds of people participate in different way in the practices. Consider for instance the practice of producing an opera which requires a variety of people who make music, sing, design costumes, do make-up, and so on, and so forth. Such variety in people and actions should be taken into account to prevent over-simplification.

Ad 3. What are the common reasons for actions in those practices? When the other involved practices are identified, the common reasons for actions of the practices can be explored. How do these aims and customs relate to human and environmental flourishing? Formulating the enquiry in terms of practices opens up the possibilities for recognizing moral reasons and techno-moral change. Focusing on actor-networks draws the attention to conflicting interests and power play, which is undeniably important, but not enough for understanding the social role of technologies. The concept of practices, however, opens the possibility to explore and compare also the morally relevant reasons for action.

Ad 4. Given these reasons and given the existing technologies, what uses are likely? When the common reasons for actions of the involved practices are better understood, the next step is to explore what kinds of uses of the technology will be likely. When people design a technological artifact, they often have some implicit ideas about the practices of the eventual user (Akrich 1992). For instance, parents as well as developers of computer games often assume that red cars are toys for boys while pink ponies are for little girls. Thus, the computer games are designed according to these presumptions: boys can race with cars and girls can brush their ponies. However, such ideas are not necessarily adequate. Having substantial knowledge about the aim and customs of the affected practices contributes to the ability to imagine the likely kinds of uses of the technology.

This step is about the relation between, on the one hand, the reasons of developers to develop, and, on the other hand, the reasons of the users to adopt a certain technology. Will people accept a new technology? And if so, for what purposes will they use it?

Ad 5. How will these uses mediate the reasons for actions in the involved practices? If step 4 explains which uses a technology will be adopted for, we can try to obtain some insights in how people's actions will be mediated by that use. Step 3 has shed some light on the reasons for actions in the involved practices and understanding how a technology might be used enables us to draw scenarios on how technologies may alter the reasons for action. Step 5 aims to obtain some understanding of how the technology can mediate the reasons for actions that are commonly adopted in the involved practices.

To readdress the example of the computer games: you will rarely find games in which boys are washing their cars and girls are horse racing, while cars need to be cleaned and horses need to run. The argument that boys naturally like cars and girls are naturally attracted to horses is hard to sustain. Through the ages, hose riding was seen as indecent for girls and hobbyhorses were typically boy toys (Nicholas 2001). So, the technologies reaffirm what we consider desirable activities and attitudes (the winner's spirit versus caring love) for boys and girls.

These five steps are formulated rather generally, as they are developed to support practitioners to structure their enquiries on a wide variety of technologies. Thus the questions need to be specified for particular cases. Chapter 7 shows how the five-step approach and the other two parts of the toolbox may be used in a policy making practice by exploring the possible future social role of intelligent driving technologies.

But first, the next sections introduce two other parts of the toolbox: that part that teaches us how to obtain answers to the five questions and the part that evaluates these answers. Section 6.3 provides a short introduction on the second part of the toolbox by presenting three existing academic approaches that specialize in studying the human-technology interaction.

6.2 Part 2 of the Toolbox: Means to Answer the Questions

How can we discuss the five steps and evaluate the social role of a new technology in practice? For this, people need tools to support imagining how technologies will mediate the three types of reasons for action (perceptions, options, and moral beliefs) in practices. How can we build fruitful scenarios about future mediation?

NEST-ethics is one of the rare accounts that aims to understand future technological, social *and* moral changes. It is a relatively young field of study, and it clearly contends with the epistemological difficulties that arise when knowledge about the future is the object of study (Swierstra et al. 2009, p. 120):

> Unfortunately, we cannot foretell the future impacts of technology. Humbled by many failed attempts in the past, we have now learned that the future is impossible to predict. Not only do we lack the necessary knowledge, but the future is essentially open and contingent on our choices, as is clear from phenomena like self-denying or self-fulfilling prophecies. Still, we cannot help preparing for it. Purposeful action assumes some degree of speculation about future impacts.

To deal with the epistemological problem, NEST-ethics works "to anticipate ... potential forms of interaction between technology and ethics" and combine these with "insights from the field of science and technology studies with notions derived from a pragmatist ethics" (Stemerding and Swierstra 2006, p. 83). In other words, to explore the moral and social mediation of technologies, NEST-ethics relies on two academic approaches: philosophy and STS (foresight TA). This section adds that in behavioral sciences several methods for studying the human-technology relationship have been developed that may provide valuable insights for further reflection. None of these approaches can imagine future technological mediation on their own, but together, they provide valuable means for drawing scenarios on technological, social and moral changes.

Reflective Tools: Philosophy of Imagination First, there is the philosophical tradition that reflects on how to stimulate moral imagination. Many authors who discuss technology stress the importance of imagining different possible outcomes and uses of technologies (Coeckelbergh 2007; Johnson 1993; Jonas 1984; Martin 2006; Pritchard 2001). Or, in Jonas' words, the "First duty of ethics of the future is visualizing the future" (Jonas 1984, p. 27). Commonly, three reasons are identified for the need of moral imagination (Coeckelbergh 2006; Johnson 1993): (1) for understanding empathically how others experience things, (2) for envisaging the full range of possibilities open to us in a particular case, and (3) for grasping what is morally relevant in a specific situation. The technological mediation of moral beliefs is another reason for moral imagination. The issue of how a technology will influence moral beliefs is relevant for imagining the future social role of technologies as well as for evaluating them.

Tools are offered for being imaginative on a moral level are among others:

- The production and study of literature, (e.g. Nussbaum 1995)
- Training in empathic world citizenship, (e.g. Nussbaum 1997)
- Thinking behind the veil of ignorance, (e.g. Rawls 1999)
- The use of metaphors, (e.g. Johnson 1993)
- The study of utopias, (e.g. More 2005)

These approaches share a basic element: they invite individuals to sit down and use their imaginations to come up with ideas on how a certain action might affect others. Although this is also what it comes down to if we want to explore the possible future social role of technology (as said, we do not have a crystal ball), it is hard to ask people to be imaginative without offering them ways to promote imagination.

Mark Coeckelbergh argues that moral imagination is frustrated because people are unable to transcend the roles they fulfill in compartmentalized societies (Coeckelbergh 2006). MacIntyre nevertheless uses compartmentalization as a tool to support imagination. According to MacIntyre, understanding yourself as an individual who can take responsibility means that you adapt a unique combination of roles. You can use the experience and views of the different role practices you participate in to broaden your moral scope.

People can use the experience of their different roles as a starting point for reflection. In reflecting on the common morality, for instance, MacIntyre argues that people can make use of experiences in their private life. People have different social surroundings and act in different practices. Attitudes toward certain situations vary with the social role adopted. By occupying different roles, people can develop a broader perspective for evaluating and adjusting their actions. For instance, an internet advertiser wants to make sure that people of all ages regularly visit the sites where his advertisements are displayed. At the same time, he may be a father who wishes his teenage children do not spend too much time on the internet and wants prevent his children to visit sites displaying adult adverts. By changing viewpoints, a broader moral imagination can be obtained.

Additionally, experiences with existing technologies can be used as analogies (Coeckelbergh 2006). For instance, Anti-Lock Braking Systems (ABS) on trucks did not increase the safety because truck drivers quickly realized that their brake path had become shorter. Thus, they brake later, and keep a shorter distance between their truck and the traffic ahead of them (Bleukx and Tampère 2004). They also took more risks. Consequently, the widespread introduction of ABS failed to contribute to road safety (for a more detailed case study on road safety, see the following chapter). Having learned from this experience, people studying road safety were quick to draw the analogy when advising industry and governments (Hoetink 2003a).

Intersubjective Tools: Technology Assessment (TA) as an Example Thus, we can become more imaginative by using analogies and by using the compartmentalization of our lives. However, new technologies still have unexpected aspects, and we do not participate actively in all practices. We cannot experience involvement in all practices, let alone in future practices that are mediated by current technologies. Beside, moral imagination is not enough: sociological imagination that enables us to understand how individual agency relates to social developments is important too (Mills 1959).

As a solution to these problems, NEST-ethics points out that STS studies aim to understand the human-technology interaction. Swierstra and Rip mainly refer to Constructive Technology Assessment (CTA) for developing techno-moral scenarios (Swierstra and Rip 2007). This echoes Verbeek's argument that mediation of perception can be investigated by CTA (Verbeek 2006). CTA simultaneously aims to study and shape the integration of science and technology in society (Rip et al. 1995, p. 5). CTA scholars do so by organizing intersubjective activities such as strategic conferences, consensus conferences, dialogue workshops, interviews, and social experiments (like role-playing) (Schot and Rip 1997).

These methods are helpful for generating a mutual understanding of current technological developments. Often distinct parties hold different ideas on what a technology is capable of and how to evaluate it. CTA helps form shared beliefs on technological issues and maps the interests of involved actors. But such a tool has more to offer: CTA can create insights about perceptions (Verbeek 2006), options (Rip et al. 1995) and moral beliefs (Stemerding and Swierstra 2006) of the involved practices.

CTA can thus enhance technological understanding, and it can help to explore the current different interests and moral beliefs. But, CTA was not developed for exploring future techno-social and techno-moral changes. Their methods for supporting power-free dialogue were designed to enable weaker actors to express their interests (which is undeniably important). Consequently, these intersubjective approaches sometimes fail to lead to trustworthy insights on human-technology relationships.

These dialogue-based methods may not close the epistemological gap. Participants of such projects may act differently to how they imagine: most people do not imagine that their moral position may change. For example, when the first genetic test for Huntington disease was under development, many people who were at risk for this neurological disease were asked whether they wanted to be tested. The genetic test would reveal to young adults whether they would develop the deadly disease or not. Most people said that they would like to make use of the test, because they wanted to be prepared for the future. However, when the test became available only one patient out of ten asked to be tested (Burgh 1997; Tibben et al. 1997).

Initially, the main worry about genetic tests was that testing yourself may also reveal genetic information of other family members. For instance, if an 18 year old whose grandfather died of Huntington wants to be tested, a positive result implies that one of the parents is also affected with the disease. This complication played an important role in early discussions. But, as soon as the tests became available, people feared the results, and it was realized (due to experiences in other families) that the tests could lead to a deep divide within families, notably between affected and non-affected members. Moreover, it turned out that people whose tests were negative suffered from substantial psychological problems (depressions). They also had feelings of guilt toward other family members (Tibben 2000). The complex psychological and moral considerations of the Huntington case illustrate that, although they can provide important information, TA methods are not sufficient to imagine what the social role of technologies will be.

Furthermore, translating the outcomes of TA into the design of technologies is hard, as the following quote on Constructive Technology Assessment (CTA) explains (Schot and Rip 1997, p. 255):

> Full-fledged CTA activities in the Netherlands and Denmark, as well as some of the recent technology policy declarations, emphasize dialogue and the articulation of demand and acceptability. This is an important and necessary part of CTA. It leads to (diffuse) societal agenda building for new technology. However, it turns out to be difficult to organize feedback of such societal agendas into actual technological development. There often is no special effort at feedback other than publicizing results and hoping that technology actors will respond.

Translating the results of the discussions into concrete designs is difficult because people do not really handle or use the technologies. CTA methods are incredibly valuable for generating mutual understanding, but they do not inform how technologies will mediate our acts. Some schools within behavioral sciences are specialized in generating insights into such user-technology mediation. Their insights can complement CTA.

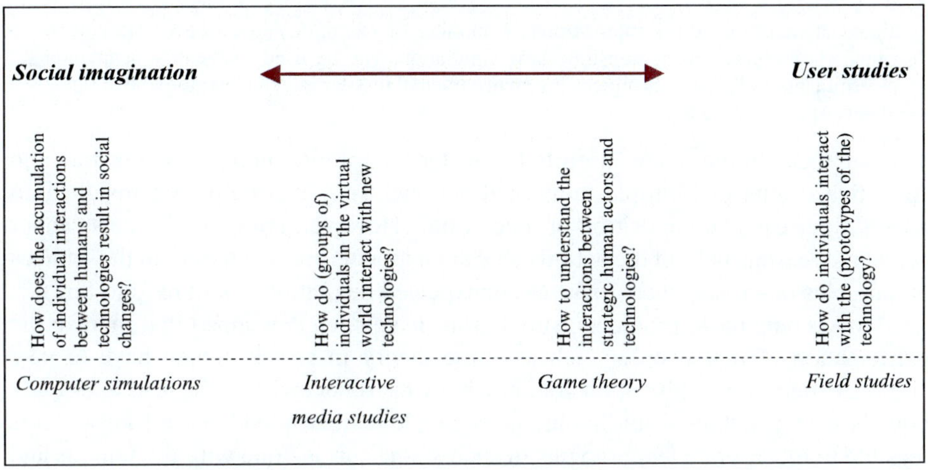

Fig. 6.1 schematic representation of the scale of methods of behavioral sciences that study the human-technology interaction

Research Tools: Behavioral Studies The five steps also consist of empirical questions on human actions, and so we can use the empirical methods of behavioral sciences developed specifically to explore the human-technology interaction of inventions. In the last few decades of the twentieth century, many behavioral sciences (such as ergonomics, cognitive sciences, and psychology) have developed methods to study how actions are influenced by new technologies. The results of such studies can be employed for supporting or stimulating our imagination.

Behavioral studies on how users interact with technologies vary in aim. To structure this variety, one can picture a scale on which the methods can be placed. The left end presents studies on how the behavior of individual users may result in social changes. The right side of the scale depicts studies on details of interaction between individual users and the prototypes of a technology. At the one end, studies address social imagination, at the other end the results of the studies can inspire moral imagination. Generally, for obtaining insights into collective actions, computer models are used. Individual users are commonly studied in controlled environments such as laboratories, although field studies are carried out as well. Schematically, such a scale can be presented as shown in Fig. 6.1.

The one end of the spectrum is based on models of large-scale social interaction that are translatable to computer simulations. Such simulations are used for all kinds of studies, from policy assessments and simulating sustainable behavior (Bossel 2000) to the study of human behavior in military situations (Pew and Mavor 1997). A recent example is the use of computer simulations to predict human behavior in relation to environmental resources (Jager 2007, p. 97):

> Computer simulation allows for the experimental study of dynamic interactions between human behavior and complex environmental systems. Behavioral determinants and processes as identified in social-scientific theory may be formalized in simulated agents to obtain a better understanding of man–environment interactions and of policy measures

aimed at managing these interactions. A number of exemplary agent-based simulation studies is discussed to demonstrate how simulations can be used to identify behavioral determinants and processes underlying environmental problems, and to explore the possible effects of policy strategies.

Computer simulation is useful, for instance, for producing insights into the collective results of multiple individual actions. They can help gain insight into what Latour calls "composition of mediation". However, computer models cannot reveal the reasons behind the actions, and such models are still based on the choices of variables that researchers make (so unexpected mediation is still hard to find).

To overcome these problems, simulations have been developed that involve the participation of real people, such as game theory in which people have to take "rational" decisions while often not knowing what others will do. Researchers study how these interactions result in outcomes people neither intended nor foresaw. It is feasible to incorporate technologies in such studies, as is done with the famous and realistic simulation game Fish Banks developed by Dennis Meadows.

The virtual environment Second Life—in which people interact with each other by means of animated avatars—is also used by behavioral scientists to see how people respond to certain extreme situations. For instance, computer animation researchers have recently started a 3-year project linking animation to psychology (Bournemouth University 2008) in order to study the "bystander effect" on Second Life. The "bystander effect" is that the more people witness an emergency, the less likely it is that an individual bystander will intervene. Studying the bystander effect in real life is difficult. Similarly, Second Life may be used to see how technologies may mediate human behavior.

Simulations involving real people enable researchers to take the complexity of human actions into account. However, many of those methods only study action outcomes, while it is essential to understand the reasons for actions if one wants to understand more fully how humans and technologies interact. This might be partially solved by incorporating studies on the social role of technologies into existing field tests. Prototypes of new technologies are tested in the field by offering them groups of users (consumers, industry, and businesses) under either controlled circumstances or in real-life. Such studies generally focus on safety issues and are compulsory in most countries (for example under the USA Consumer Product Safety Improvement Act). If the testing of new products is mandatory, why not also incorporate other aspects of testing?

However, not all technologies can be tested with prototypes and a technology is likely to fulfill a different social role when introduced on a large-scale. Consider, for instance, gender selection, which is nowadays only granted for specific medical reasons (like the prevalence of severe gender related diseases). One cannot simply foresee the social impact of legalizing gender selection, by just extrapolating the effects of medically initiated gender selection. Nor is it possible to launch such an experiment in Second Life. So in such cases behavioral science may not result in many insights.

So, we can conclude that each of the three approaches (imaginative philosophy, TA, and behavioral studies) has both strong and weak points, providing some valuable insights, while obscuring other elements. There is no all-encompassing, epistemologically satisfying method to foretell what the social role of a technology will be. However, combined, the approaches provide substantial means to make more educated guesses on what possible techno-social and techno-moral scenarios might be. In other words, we have some tools that can provide inspiration for drawing up these scenarios. Using not just one, but a selection of such tools may provide us with some clues on the likelihood of these scenarios.

6.3 Part 3 of the Toolbox: Evaluating the Social Role of Technologies

So far, the focus has been on the descriptive tools for taking responsibility. But taking responsibility also entails making decisions. Thus, the toolbox needs a third kind of tool to facilitate the moral evaluation of the expected social roles of technologies. Which social role is to be preferred over others? And which social roles are undesirable?

Moral evaluations are based on moral beliefs, which differ across practices and evolve over time. Consequently, the evaluative tools pose a problem: should we focus on current moral beliefs and judge the technologies from this perspective? Or, should we evaluate a technology on the basis of moral beliefs of another involved practice? This question is not just theoretical, as Dutch agricultural developments of the twentieth century illustrate.

After the Second World War, a commonly shared priority was to make sure that the Netherlands would never again suffer famine. Important players in the field (the government, banks, insurance companies and the food industry) worked to further industrialize agriculture. Many new technologies were applied to increase food production until the surplus of food grew beyond both national and export needs. From the 1970s to the 1990s, huge amounts of food (mainly dairy products, but also meat) were continuously being destroyed, while the national government and the European Commission continued to stimulate the increase of food production by subsidizing farmers for their surplus. Breeding techniques, antibiotics and technologies like milking machines were employed to increase yield.

However, due to continued industrialization, fewer and fewer people were employed by the farms. Most people no longer had direct experience with or knowledge of food production. The only relationship many people had with other animals was the caring relationship with their pets. Pets became members of the family: they lived under the same roof as their owners, receiving much individual care. Special relationships can grow within such a setting: the individuality, the cleverness and the "cuteness" of these animals become more prominently addressed. At the same time, the bioindustry began to treat their animals more and more as exploitable production units (Waelbers et al. 2004).

In the 1990s, the outbreaks of foot and mouth disease and swine fever led to the destruction of thousands and thousands of animals, and people were bluntly confronted with the high-technological and animal unfriendly aspects of the industry. Many people responded with disbelief and deep disapproval when they saw on television how excavators picked up many dead animals at once and dumped hundreds of them in huge containers. A mediating effect of the increased industrialization of agriculture was that far less people became involved in food production, and so people started to assign a different moral status to animals. They criticized the practice of animals being treated only as means, as things (Waelbers et al. 2004). Now, they judged the decision to ensure food abundance as being immoral toward animals and the environment.

But how are we to judge future social roles if moral beliefs shift over time? We can start with formulating some pragmatic evaluation criteria. Pragmatic evaluation criteria are based on the idea that possible undesirable social roles of technologies should be either restorable or outweighed by the desired social role. Put differently, the new social role should not create bigger problems than the problems which the development or employment of the technology intended to solve (Tenner 1997). Furthermore, commonly a scenario is more desirable if it leaves more choices open in the future. It is fair to allow future people to make their own choices. We are not sure about the future social role of new technologies. Therefore, it is often desirable to preserve enough options, so we can decide with more knowledge in the future.

Consider the following case in which this last criterion played an important role. To improve the economy of a Dutch rural region, a one-way railway bridge was built over a river about a century ago. Freights could run between the south of the Netherlands and the Ruhr area in Germany, and people could commute between two Dutch provinces. The new railway along with several other initiatives, promoted economic growth. However, the bridge had only one set of tracks, so only one train could cross the river at a time and today has become an obstacle to further economic growth. It often increases delays because the capacity of one train every 5 min is no longer sufficient. Demolishing the old bridge and building a new one is undesirable because the trains would not be able to run for a long time.

So, the decision to build a one-way or two-way bridge had consequences for literally hundreds of years. But fortunately, the architect of this bridge took his long-term responsibility seriously and foresaw that eventually (after several generations) a one-way bridge would perhaps not be sufficient. He had designed pillars and foundations strong enough to carry a two-track bridge. In addition, he built the upper part of the bridge so that it could be replaced relatively easily.

Pragmatic evaluation criteria are important starting points for evaluation, but they do not address the full scope of relevant moral considerations. Substantial moral criteria (norms and values) need to be specified. What moral values should we prioritize for evaluating the social role of technologies: the current morality of the practice in which the technology is developed, or the morality of the other involved practices (including possible future moral beliefs)? Current moral beliefs cannot simply serve as an independent or objective basis for evaluating the social role of technologies since these beliefs themselves may be technologically mediated (Swierstra 2009).

How can these insights be acknowledged without falling into the trap of relativism? We can find a basis for the second type of evaluation criteria in the following: the aim of developing technologies should is often to serve other people. So the evaluative question is: How are other people served best? The most comprehensive reply is that people are served best when their flourishing as well as the flourishing of the environment is promoted.

Which substantial criteria can be used to see whether the social role of technologies promotes human and environmental flourishing? MacIntyre's four elements of a good life (MacIntyre 1999) (see also Sect. 4.4) can be translated for this aim. Such criteria are not moral principles: they do not tell us to "do this" or "not do that". MacIntyre's criteria inform us about what we need for a flourishing life. First, we need to fulfill our bodily needs. Secondly, we need to be able to acquire skills and techniques. Third, we need to be able to socialize with others and fulfill a meaningful social role. And, fourth, we need to be able to develop ourselves as good persons. To translate these criteria for evaluating the social role of technologies in the sense that one has to have good reasons to develop or employ a technology that frustrates any of these four criteria. However, to see whether a technology frustrates these criteria, they need further specification, which can only be done intersubjectively in context. Consider the four elements in more detail.

Bodily Needs MacIntyre's first element of flourishing is the requirement to fulfill bodily needs. We cannot flourish in a world in which there is not enough food, water, housing or clean air. So far, the criterion can be rather straightforwardly translated. But technologies mediate what we consider good for our bodies. The microscope mediated our perceptions of hygiene and diseases (Barnes 2006): the discovery of micro-organisms enabled a better protection of health. But as our bodies have become better protected against all kinds of micro-organisms, we have also become more vulnerable to others such as salmonella.

Technologies mediate what our bodies need. Recall Tenner's example of shoes causing your feet to become soft and in need for protection (see Sect. 5.3), made this quite clear. On the one hand, technologies should be evaluated with regard to the protection of our biological vulnerabilities and the fulfillment of our bodily needs. On the other hand, these bodily needs and biological vulnerabilities are contextual. Techno-social changes influence our needs and vulnerabilities. Therefore, applying this criterion requires an intersubjective and imaginative approach for interpreting and evaluating technological mediation.

Skills and Techniques MacIntyre's second element refers to possessing skills or techniques. Here again, technologies mediate skills and techniques. New skills and techniques may arise. For instance, robot assistance enables surgeons to perform subtle brain surgery that is unthinkable to perform by hand. Other skills may alter or disappear. Tenner describes how writing skills have been altered due to ballpoint pens, typewriters and computers (Tenner 2004). Students often complain that writing 3 h during an exam is physically too demanding. They are also much slower at handwriting than students were 10 or 15 years ago, because they rarely write anything by hand: all papers and assignments have to be typed.

The second element of good should not only be translated into the question of whether a technology facilitates skills and techniques in a desirable way. The skills and techniques may be mediated as well, and thus we need to imagine and evaluate the question of whether a mediation is desirable. Do we lose some skills and techniques that we think are valuable in themselves (such as handwriting) or does the new technology enable us to gain new valuable skills?

Social Roles MacIntyre's third element is that social roles should be worth pursuing. Following the Luddites, many philosophers and sociologists of technology have argued that the traditional crafts, which were highly appreciated in society and rewarding to fulfill, have been transformed into uninteresting, unskilled labor. However, technologies also made some practices more interesting: for example constructive technologies opened many interesting possibilities for architectures and new media offers many new possibilities for artistic expression. New technologies can also emancipate people, providing them with the opportunity to fulfill valuable social roles, as the example of *Second Life* made clear.

A responsible introduction of new technologies asks for imagining and evaluating how that technology will enable people to fulfill valuable social roles.

Morality MacIntyre's last element is that you should be able to become a virtuous person. The social role of a technology can be evaluated for its potential to support people's abilities to do and be good. The previous chapter argued that at least a limited amount of freedom is needed to live and act according to one's own moral beliefs. Since this is such a central characteristic of what it means to be a human actor, one should have quite strong arguments for developing technologies that limit such freedom. But technologies do not only limit options, they also create new possibilities ones and thus offer new freedoms. The previous chapters also discussed how we can evaluate our moral beliefs by using the capacity of practical reasoning and imagination. As long as we have these capabilities, we are able to be virtuous.

Thus, this fourth element can be used as a starting point for evaluating whether the social role of technologies is desirable by questioning whether people will be able to live a virtuous life. But here again, it is important to recognize that technologies continue to mediate the evaluative criteria. What we consider virtuous alters over time, as the example of the appreciated student's attitudes and the design of the classrooms illustrates. Such mediation of moral beliefs makes it hard to use this element as a criterion: what if we end with a possible social role in which being detached or being hateful is a virtue? Does the technology support the freedom required for being a virtuous person?

This translation of MacIntyre's four elements leads to general criteria for evaluation. These criteria can be summarized with the question of whether the social role of technology enables people to meet bodily needs, carry out tasks, fulfill valuable social roles, and develop virtues. In this sense, the four meanings of good in relation to human flourishing can be used as criteria. However, these criteria still need interpretation regarding the specific technologies and their mediating role. Furthermore, when conflicting values are found, they need to be weighed against each other. Such interpretation and the final evaluation can only be made intersubjectively (see also Sect. 6.3).

MacIntyre's four elements can serve as a starting point for discussion on the desirability of the possible future social roles of technologies. Exploring and evaluating the social role is an intersubjective endeavor, and thus, the evaluation criteria do not have to deliver easy answers. In the organized communication, the involved practices can provide insights for selecting, interpreting and complementing the criteria to make them suitable for discussing concrete technologies. They have to decide what is most valuable by selecting the essential points for the specific evaluation, to decide how to interpret these points, and to see how these points may alter over time.

Summarizing, this chapter developed a toolbox to help practitioners to take responsibility for the social role of technologies. The first part offered a five-step approach to structure the deliberation on the future social role of technologies by pushing forward five questions. The second part, explained how a combination of reflective, intersubjective and empirical methods may help to imagine what the possible social roles might be. The third part of the toolbox provided some guidelines for evaluating the possible social role of technologies. To picture how this toolbox can work in practice, Chap. 7 shows how it may be used for a responsible introduction of "intelligent cars".

Chapter 7
Case Study: Taking Responsibility for Future Driving

How can we take responsibility in practice? The previous chapter introduced a toolbox that supports imagining the future social role of technologies: five questions were offered to structure imaginative reflection, some examples of reflective, intersubjective and empirical (behavioral) research methods were explained to support the desired imagination, and some starting points for the collective evaluation of the possible social role of technologies were presented. However, these tools aim to be widely applicable, and, consequently, they need to be clarified and specified for real-life cases. Clarifying and specifying has to be done intersubjectively to obtain a comprehensive and well prioritized evaluation (meaning organizing strategic conferences, consensus conferences, dialogue workshops, and interviews). This chapter explains briefly how the toolbox can generate questions for an intersubjective study on the intelligent car of the future. In other words, the aim of this chapter is to show how applying the toolbox can generate new insights and questions on expected and unexpected social impacts.

Note, the aim is not to judge current roads, cars, traffic regulations or driving styles. Nor is it the aim to provide a full analysis and evaluation of future driving technologies, because such an analysis and evaluation requires the participation of the involved practices. The aim of this chapter is to show how the toolbox can support the enquiry into the possible future social role of technologies. For this, first it needs to be explained which technologies are needed to realize the "intelligent car of the future", and how the social role of these technologies has been studied so far.

7.1 Studying the Social Role of Intelligent Cars

Autonomous Driver Assistance Systems If we are to believe the European Commission (2010), the intelligent car of the future will have many integrated Autonomous Driver Assistance Systems (ADAS)—also called Intelligent Vehicle

K. Waelbers, *Doing Good with Technologies: Taking Responsibility for the Social Role of Emerging Technologies*, Philosophy of Engineering and Technology 4,
DOI 10.1007/978-94-007-1640-7_7, © Springer Science+Business Media B.V. 2011

Systems (IVS)—that will almost fully automate driving. Current ADAS technologies are intelligent systems that support more comfortable and safer driving. The most well-known ADAS technology is Autonomous Cruise Control (ACC). Autonomous Cruise Control systems are also called Advanced Cruise Control systems, Adaptive Cruise Control Systems or Intelligent Cruise Control systems (ICC). ACC systems uphold "a previously set driving speed, which is automatically reduced when the headway time—a measure for the distance between two vehicles, which is commonly programmed between 1 and 3 s—is shorter than a pre-set value" (Eenink 2009, p. 5). The desired speed is programmed manually just as in ordinary cruise control systems, but it also measures the headway time with infrared, radar or object recognition. If the headway time becomes shorter than the installed minimal value (between 1 and 3 s), the system slows down by reducing speed, changing gears or by using the brakes (depending on the brand). So far, few of these still expensive systems have been sold, but new variations are being developed. These new variations are not only cheaper, but also combine various ADAS technologies, such as Forward Collision Warning, Lane Departure Warning Assistance, Directional Control, Intelligent speed adaptation, Blind Spot Detection, Car2car communication, Driver Drowsiness Monitoring and Warning, Dynamic Traffic Management, Intersection Assistant, Pedestrian Protection System, and Local Danger Warning. Table 7.1 lists some examples of ADAS technologies.

All these technologies are information technology based solutions, which are currently separately developed with the eventual aim of creating an integrated system for what is sometimes called "the Intelligent Car of the future". Such a car will have a large amount of autonomy and will change the practice of driving substantially. The European Commission has launched an extensive project called "Intelligent Car Initiative" to promote the development of a fully automatic car. They support a long (and still growing) list of ADAS technologies, for they argue (European Commission 2010):

> Imagine a world where cars don't crash, where congestion is drastically reduced and where your car is energy efficient and pollutes less. Today Information and Communications Technologies (ICT) are starting to make this dream true. Your car is becoming smarter, helping to reduce Europe's road transport problems. … We all depend heavily on transport in our everyday lives. However, ever-increasing road traffic generates serious social problems: congestion of road networks and urban areas, damage to the environment and to public health, energy waste and, above all, accidents. 40,000 people die every year on Europe's roads and many more are injured.
>
> Fortunately, advanced information and communication technologies (ICTs) can now be incorporated into onboard "Intelligent vehicle systems", offering new solutions to today's transport problems. These high-tech systems have great potential to:
>
> - help drivers prevent or avoid traffic accidents
> - mitigate the consequences of accidents that do occur
> - provide drivers with real time information about traffic on road networks, thereby avoiding congestion
> - find the most efficient routes for any journey
> - optimize engine performance, thus improving overall energy efficiency.

Table 7.1 Examples of different ADAS technologies

Abbreviation	Name	Function
FCW	Forward Collision Warning	These systems give an auditory warning when the headway distance to a preceding vehicle becomes too short or the approach speed is too high; they aim to prevent rear-end collisions (Eenink 2009)
LDWA	Lane Departure Warning Assistance	This system gives a signal when a pre-set distance from a line marking is exceeded or approached too fast. LDWA is intended to prevent single vehicle crashes in which a vehicle goes off the road (Eenink 2009)
DC	Directional Control	This system continuously compares the steering angle and the individual wheel velocities and breaks a wheel if it deviates from the correct value. DC can prevent a vehicle going into a skid or can limit its consequences (Eenink 2009)
ISA	Intelligent Speed Adaptation	ISA monitors the speed of the vehicle and compares this with the local speed limit. Some ISA systems enforce the driver to respect the speed limit, while other systems make it harder, but not impossible, to exceed the speed limit. Some systems just signal (by sound or display) the driver that the speed limit is exceeded (AVV 2001)
CWS	Obstacle and Collision Warning System	CWS detects vehicles or other obstacles on the road. Some systems just warn the driver for collisions, other systems enforce a full halt. Current solutions use short-range radar sensors, future systems will use long-range or near-range radar sensors or LIDAR and video image processing or a combination of these sensors (European Commission 2010)
VRUP	Pedestrian or Vulnerable Road User Protection	VRUP protects vulnerable road users such as pedestrians and cyclists. The sensor systems monitor the headway and distinguish vulnerable road users from other obstacles. VRUP can brake and can trigger protective structural actuators like the air bags in the bumper, lifts in the hood of the car (European Commission 2010)

However, the European Commission is too naïve in this promotion of the car of the future, and a responsible introduction of ADAS technologies needs a more balanced assessment of the possible social roles of these technologies. All these listed advantages do not address any possible undesired impacts. ADAS technologies raise many expectations, but, there are also many questions to deal with if we want to realize a responsible introduction of these technologies. All technologies fulfill a mediating role, and such a role can lead to unexpected and undesired outcomes.

Which role might the ADAS technologies fulfill? Driving is likely to alter drastically if all technologies are introduced, but what will it look like? What will their effect be on road safety and congestion? Will drivers accept responsibility if ADAS technologies take over many tasks autonomously? What will their influence be on our experience of and our attitude toward driving? And what other social impacts are to be expected?

The five-step approach is a good starting point to generate and discuss such questions in a more focused manner. The results of existing behavioral studies can be used to inform this approach.

Studying the Social Role of ADAS Technologies Existing behavioral studies with early versions of such systems can be used to support the enquiry on the possible social role of future, more advanced ADAS technologies. This chapter focuses almost exclusively on Dutch studies since there are significant differences between countries in driving styles, traffic rules, roads and climate. In the Netherlands, institutes such as the Netherlands Transport Research Centre (Adviesdienst Verkeer en Vervoer, AVV) and the Dutch Institute for Road Safety Research (Stichting Wetenschappelijk Onderzoek Verkeersveiligheid, SWOV) study the safety issues of such echnologies with multiple empirical and intersubjective tools. These institutes work to obtain insights into the possible interaction between ADAS technologies and the user prior to the widespread introduction of these technologies into society. This knowledge is used to support the car industry (for example, Oei 2003), to advise policy makers (for example, Tabula Rasa 2004), and to inform users (for example, Tonnon and Jagt 2003).

In their studies, road safety institutes explicitly acknowledge that many safety measures have what Tenner calls "revenge effects". When car safety technologies are employed, road safety can decrease. As the safety of the car is increased, drivers will show more risk-taking behavior. The introduction of ABS, for instance, lessened the number of multiple collisions, but (and this is an important but) far more vehicles crashed into trees and bollards because drivers trusted that they could brake later (Mazzae and Forkenbrock 2003). To explain the idea that safer cars lead to more risk-taking behavior, Top Gear presenter Jeremy Clarkson—presented with an honouree PhD in engineering by the School of Technology of the Oxford Brooks University and the University of Brunel in 2005—once argued that people would drive more safely if, upon impact, an iron spear instead of an airbag would pop out of the steering wheel.

Having learned from the experiences with safety technologies such as ABS, the SWOV and AVV started to study these kinds of technological mediation in practice (the concept of mediation is not commonly used in the field of care safety studies, but the mechanisms are. I have taken the liberty to translate their studies in terms of mediation). Therefore, they increasingly conduct empirical experiments for studying the human-technology interaction. As argued in Sect. 6.3, behavioral studies on human-technology interaction can be placed on a scale varying from computational models on social change to studies that observe individuals handling prototypes. The same scale is also recognizable in road studies. Roughly, research institutes use three types of empirical research methods to study the human-technology interaction (Hoetink 2003a): (1) computer models (also called traffic simulators), (2) driving simulators, and (3) field tests on ranges and roads. Let me focus on the last two methods.

Driving simulators are used to study driver's behavior. Most car simulators consists of a cabin in which the interior of a car is rebuilt, including gas, clutch, brake pedals, force-feedback steering wheel, and stalks. Multiple screens and speakers provide the drivers with three-dimensional audio-visual input. The behavior of drivers can be measured by their performances in the simulation game, observation cameras, and with interviews or surveys. "Driving simulation scenes are comprised of the layout of the roadways, shape and appearance of buildings and other cultural features, vegetation, the shape and appearance of synthetic vehicles, and the properties of drivable surfaces. Scenarios involve the behavior of the synthetic vehicles, traffic control devices and variations of weather and lighting" (Stoner et al. 1997, p. 18).

Field tests are organized on test tracks and on roads. Test tracks explore how humans and technologies perform together under preset conditions. Because of increasingly advanced computer simulations, test tracks may become less favored (Gietelink et al. 2004), but nowadays test tracks are still important for safety tests as well as in the development of new technologies. On a test track, the systems are put to further trial to see how they respond to weather conditions like fog and rain, and how they function during special manoeuvres like acceleration and taking different kinds of bends. Tests on the roads are employed to examine how multiple humans and technologies perform together in daily practice. Sometimes, temporarily uninhabited neighborhoods are assigned for these tests, see for example the first Dutch experiment with factory installed ACC. This test was conducted by the SWOV in 2003, using a Nissan Primeras (Oei 2003). The Nissan Primera was driven by ten different people from and recruited by the SWOV. The subjects had read the manual on how to operate the ACC. In total, they drove 1,700 km.

These three types of tests mainly focus on safety issues, while some also measure fuel consumption. Sometimes these studies are supplemented with preference studies (questionnaires). On the one hand, these studies show that ADAS technologies are promising, but on the other hand, they also show some unexpected results. In 2003, the Dutch Institute for Road Safety Research (Stichting Wetenschappelijk Onderzoek Verkeersveiligheid, SWOV) published a press release stating the following (Tonnon and Jagt 2003): "Motorists need to be informed about the safe use of Advanced Cruise Control (ACC). Research of the Institute for Road Safety Research SWOV has shown this is necessary. If used wrongly, ACC can result in more crashes, even though it increases driving comfort and can contribute to safety". The press release warns us that ACC only increases safety "on motorway journeys if there is not much traffic and if the view and weather are good". On all other kinds of roads, on busy motorways and in different weather conditions, the early ACC systems can decrease safety. The press release was based on the extensive literature research from existing behavioral studies publications. The report summarized positive as well as negative outcomes of the human-technology interaction (Hoetink 2003a, b). The positive outcomes are:

- The average speed outside rush hours decreased.
- The systems make driving less tiring and more comfortable.
- The percentage of very short headway times decreases.
- The average fuel consumption decreased 3–10%.
- The driving speed was more constant.

However, also noteworthy negative outcomes were found:

- In heavy traffic the average speed increased.
- Road use changed: people switched lanes more often or tended to drive for longer times on the left (fast) lane of the motorways to prevent the system from braking for cars in front of them (in the Netherlands, the upper left lane is intended for overtaking. It is finable to drive on the upper left lane while not overtaking other cars).
- When the system was used on secondary roads, the headway time decreased, more unsafe overtaking actions took place and people responded slower to right of way situations.
- People often reacted inadequately in critical situations as the drivers were less alert.
- Behavior of ACC users can be difficult to predict by non-ACC users.
- The use of ACC in bad weather conditions (fog or rain) may lead to failure of the system.
- Pedestrians, cycles and mopeds are not easily detected by ACC systems, and drivers show more asocial behavior toward these groups.

The results of such behavioral studies can support the imagination as they help to identify possible forms of mediation. But, these behavioral studies are not developed to study the wide variety of social impacts of ADAS technologies.

First, these results were obtained from studying early versions of Advanced Driver Assistance Systems (ADAS): more developed systems may sort different results. Secondly, these studies focus on one type (or sometimes a combination of two or three) of ADAS technology. However, the differences between the various systems, the interaction between multiple ADAS technologies, and the accumulative effect of several systems on the driver should not be underestimated. Third, these studies mainly focus on safety impacts and eventual revenge effects on safety. Other aspects like experience, comfort, fuel consumption and status need to be considered too.

As argued previously, to take responsibility for the social role of technologies, being imaginative is crucial. Thus, this chapter elaborates on behavioral road safety studies and the scarce intersubjective data using some tools of imaginative philosophy such as drawing analogies, making use of compartmentalization and using cultural sources such as television programs. The following sections discuss how these approaches can be combined to discuss the five steps. To recall them:

1. What are the aims of the technology? (Sect. 7.2)
2. Which practices will be affected? (Sect. 7.3)

3. What are the common reasons for actions in those practices? (Sect. 7.4)
4. Given these reasons and given the existing technologies, what uses are likely? (Sect. 7.5)
5. How will these uses mediate the reasons for actions in the involved practices? (Sect. 7.6)

Since road safety studies have been working for several decennia to understand the human-technology interaction on the road and in the car, the amount of information they have produced is enough to fill various books. It is impossible to be all encompassing in this chapter. But, since it is not the aim of this chapter to deliver a complete and detailed scenario about the future social role of all ADAS technologies, I do not consider this to be a problem. This chapter aims to show how the five steps and a combination of reflective, intersubjective and behavioral research tools may be useful for the enquiry in real-life cases. For this aim, presenting a selection of the road studies will do. Since driving styles, traffic rules, roads and climate differ significantly between countries, this selection will consist of mainly Dutch studies.

Section 7.7 explains more about evaluating the results of these five steps.

7.2 Step 1: The Aims of the Technologies

The first step is to find out what the aims of the ADAS technologies are. As argued in Sect. 6.2, technologies are made and used for a reason, and a relevant question for taking responsibility is whether the aim supports human and environmental flourishing. For answering this question, we should not only understand the primary aims (referring to the functionality of the technology) but also secondary aims. ADAS technologies are developed with the primary aims of improving driving comfort (Hoetink 2003b), and increasing safety by preventing collisions (Hara 2004). So, we may start with noting that the ideals of the European Commission nicely fit the reasons why the car industry developed these technologies, and these ideals support human flourishing. Still, comfort and safety are not the only two aims of the car industry to develop these kinds of new technologies. The industry also refers to the positive results (the secondary aims) for the environment (reduction of fuel consumption) and for people (preventing congestions).

So, four aims are mentioned for developing ADAS technologies: increasing safety and comfort, and decreasing congestion and fuel consumption. But the explicit aims do not have to be the only aims or the most important ones. To find out what the other aims are, it is important to interview the developers and managers of the technology. When it comes to ADAS technologies, an imaginable answer is that manufacturers try to be the first to introduce a new technology or a substantially improved version of the technology. Many manufacturers develop new technologies to be the first and most high-tech branch. Toyota—the first manufacturer that offered "radar cruise control" (Treece 1997)—claims that it offers nowadays "the world's technologically most advanced car". In their television commercials, they suggest a clear connection between driving their high-tech Prius and being a smart, trendsetting city-dweller. For them the aim of developing a high-tech car is also to create a "young-and-dynamic" image.

The question is whether these aims will be realized. New technologies may suffer from revenge effects in the sense that the human-technology interaction will actually lead to the opposite results. Furthermore, the social role of technologies can create effects that are not aimed for. To explore such possibilities, it first has to become clear which practices will be involved.

7.3 Step 2: The Involved Practices

Even when the intention is to support human and environmental flourishing, unexpected social roles can still pose problems. To acquire some ideas about the social role of technologies, the aim of the second step is to explore which practices might be affected.

Remember that practices have a slightly different meaning here: The focus is not on activities that aim to produce intrinsic goods such as producing art, music, and knowledge. For me, a practice is also a common cooperative human activity that requires virtues and that involves generally shared expectations. Thus, being alone, watching television is not a practice; neither is picking one's nose. But traffic participation is a practice: it fulfills the three criteria of the given definition: First, it is a human activity that requires cooperation. When driving a car, you need to adapt your behavior to the behavior of other road users, and you need to communicate your plans to others. Secondly, the shared expectations are important: unexpected behavior is dangerous because people need to anticipate each other's actions on the roads. Third, for safe driving you need to behave virtuously: careless, impatient, disrespectful attitudes are undesirable.

It is hard to foresee which practices will be affected. Intuitively, people tend to select practices based on the aims of the development and management of technologies. Therefore, road safety studies focus on traffic participation. But other practices are relevant as well. For instance, what does it mean for training youngsters if the car of the future becomes reality? Do we need new traffic regulations if driving alters due to ADAS technologies? Will we need to redesign our roads and traffic signs to enable the efficiency of the ADAS technologies? And how will so many IT based technologies change the jobs of car mechanics? The wide introduction of new ADAS technologies will affect many practices such as training (young) people, regulating traffic, planning roads, and maintaining cars. And because it influences so many practices, it is difficult to anticipate its social role.

The more practices we involve in the enquiry, the more insights will be obtained in the social role. When discussing a technology that is deeply embedded in our societies (like driving technologies) the list of involved practices can become incredibly long. However, involving too many practices also makes the enquiry too complex (or diffuse) to be manageable, and time-consuming while often in daily practice choices have to made quickly. Therefore, to remain practicable, a reasonable selection has to be made. Together, they can for instance decide that the likely effect on spatial planning will be minor and less relevant to the evaluation than the training of new drivers. To explain the next three steps, the focus is here on three practices: (a) participating in traffic, (b) training for traffic participation, and (c) regulating traffic.

When explaining in the next steps how ADAS technologies might mediate these practices, the diversity within the practices should not be neglected, see the example of the MacIntyrean practice of the production of an opera which requires various people who make music, sing, design costumes, do make-up, and so on, and so forth. Similarly, the practice of participating in traffic is plural, and understanding the diversity within a practice is essential for obtaining insight in the social roles of ADAS technologies. In for instance interviews and discussions, people participating in the involved practice can inform the enquiry about existing diversities.

For instance, when it comes to the practice of participating in traffic, professional driving such as driving taxis, buses and trucks, is in some respects distinct from non-professional driving. Of course they have to commit to the same traffic regulations, but some of their reasons for actions may differ, as we will see in Sect. 7.4. The main focus of the ADAS studies has long been on passenger cars and mainly non-professional driving. But, professional drivers might also employ the technology, with different results, as a limited study showed (Eenink 2009). Next, participating in traffic does not only involve driving: other road uses such as walking and cycling are important too. In safety studies, road users like pedestrians are often presented as passive victims who need protection, but as to be argued, they can become active.

7.4 Step 3: Common Reasons of Actions of the Involved Practices

The third step draws our focus to the common reasons for actions (in short, motives) of the affected practices regarding human flourishing. Formulating the enquiry in terms of practices opens the possibilities for *also* recognizing moral reasons. This enables us to explore in later steps how the technologies mediate the practices. The question to start with is why people are involved in the practices. What do they strive for? STS tools such as in-depth interviews are well suited to find out what the reasons are and how they relate to the aims of the technologies. In this section, some results of examples of intersubjective studies are mentioned to illustrate which insights this step strives for.

7.4.1 Participating in Traffic

When thinking about the reasons of why people participate in traffic in one way rather than the other, the first thing that comes to mind is transport. Since driving is about the transport of people and goods, many drivers are motivated by the urge to be efficient. This is especially true for professional drivers, because the longer it takes to transport people or goods, the more expensive it becomes. But, as mentioned, we should consider the diversity within the practices. To understand how technologies will mediate the reasons of actions within traffic participation, it is important not to think only about transport and efficiency. Which other reasons are recognizable in participating in traffic?

Little is known about why people choose a certain means of transport, car type or driving style, and more research is recommended (Hendrickxs and Schoot-Uiterkamp 2006). The few existing studies focus mainly on non-professional drivers. One of the research projects of Tabula Rasa (an advice agency that performs research on communication and behaviour adaptation for the The Netherlands Transport Research Centre) shows that in the practice of non-professional driving, the following intrinsic motives (reasons) are seen as important by the drivers themselves (Tabula Rasa 2004): freedom, fun (including sport and comfort), safety and status.

The two main motives are freedom and fun (Tabula Rasa 2004). Freedom—the ability to go from one place to another—is seen as the most important value of driving. Furthermore, some people find driving enjoyable. This second motive connects to the first: anonymity in the car and the freedom to decide where to go is part of the joy. But people also value the opportunity to listen to one's own music and to call friends without interference of family or colleagues. The research further shows that one's own safety is clearly a less important motive for driving than freedom and fun. For many people, freedom and joy of driving are important enough to take some serious safety risks—that is, as long as one does not have passengers. To imagine the future use and mediation of ADAS technologies, it is important to understand how the technologies interfere with these two motives. Will we still be free and will driving still be fun if the autonomous car of the future becomes reality?

For many people driving is fun: on Sundays, some people just drive a few blocks or more to relax and be out of the house for a while. Fun consists of three elements: comfort, recreation and sport. As we have seen in the first step, the main aim of ADAS is to increase comfort. But will it actually be used for this aim? For driving to be fun, however, it is important that the "arousal level" is optimized (Geen 1995). If drivers have too much stimuli to which they have to respond, they will become tired and stressed. If they have too little stimuli, they will get bored and start to seek other another activities such as calling someone, speeding, eating or doing administrative work. So, if ADAS technologies reduce the arousal level too much, they might introduce new risks.

Next, interviewees mention that status (the impression you make on other people) is an important reason for driving. People of different social backgrounds mention means and forms of image building. First, people mention the type of car, but also driving style, technological gadgets, and driving behavior (for instance, stopping for pedestrians versus a sporting driving style) play an important role. Most interviewees admitted recognizing the status reasons of other people, but denied that image considerations influence their own choices (Tabula Rasa 2004).

So, there are at least four reasons for action in traffic participation that need consideration in the next steps: freedom, fun, safety and status. But these reasons are derived from interviews with non-professional drivers. To obtain a full picture, the reasons behind more vulnerable road use and professional driving need to be investigated in the same manner as has been done for non-professional driving. We cannot simply presume that other road users have the same reasons. To take vulnerable road users as an example, one might presume that safety is more important for them because these road users lack the protection of the car. But, it is difficult to see what status has to do with taking a walk. Moreover, other road users may have additional reasons for their participation in the traffic. Walking and cycling is often also seen as recreational and sporting. Such reasons may result in different road behavior. For instance, racing cyclists behave differently on the Dutch provincial roads than secondary school students or other cyclists do. Most provincial roads have special lanes for cyclists, and people who cycle to school or to their work make much use of them. Racing cyclists often cycle on the main road as this enables them to overtake each other and speed up.

7.4.2 Training for Traffic Participation

Participating in traffic is a complex practice, which requires learning many skills. People need to learn more than how to drive a car— they also need to learn the regulations, to recognize and solve dangerous situations, and to anticipate the actions of other road users. They also need to be able to perform these tasks very quickly. Thus, there are at least two reasons for training and examining (young) people for traffic participation (SWOV 2008a). First, for safe and comfortable driving, the activity has to become second nature. It is essential to incorporate the basic tasks to such an extent that you do not have to think about where the pedals and switches are. Secondly, people are trained to prepare for complex events and emergencies. Notably, recognizing dangerous situations is an important skill to learn (SWOV 2010). To be safe, people need to know what to do in a wide variety of situations.

The question is how these reasons relate to new ADAS technologies. ADAS technologies take over basic tasks, as well as the recognition of dangerous situations. So, one can argue that drivers have less skills to learn. Will driving become so easy that having a license is no longer required? But ADAS technologies also require new skills: they need to be programmed and their feedback needs interpretation. What does this mean for the practice of training for traffic participation?

7.4.3 Regulating Traffic

As the quote from the website of the European Commission made clear (see introduction), regulating ADAS technologies is about increasing safety, reducing fuel consumption and preventing congestions. So, in regulating new technologies, the

European Commission will use these reasons as a basis. But, as argued, we should not oversimplify the complexity of the involved practices. Thus, we should not be too eager to limit our enquiry to the list the website of the European Commission provided us. There are several other reasons recognizable in the daily practice of regulating traffic.

For example, governments have to deal with responsibility issues. Since ADAS technologies take over human tasks, the government has to consider how responsibilities may shift from one actor to the other. Nowadays, when cars collide, the drivers bear the responsibility, but would that still be fair if an ADAS system fails? Can a future driver that has learned to trust his supporting technologies be blamed if these technologies fail?

Furthermore, regulating traffic is not always about traffic issues. For instance, the government is involved in guaranteeing working conditions of professional drivers: safety and comfort need to be guaranteed. To control that working hours are not exceeded, the governments of many European countries have taken measures such as the introduction of a compulsory tachograph. Additionally, the Dutch highway patrol checks whether there are drivers on the road with tax debts. They also check whether you are using your company car for private purposes, without paying income tax on this benefit in kind. Knowing this, questions arise about the possible dual use of ADAS technologies by the government.

Here, only a few preliminary indications of the reasons in the practices are provided. As argued, to understand the full diversity within the practices intersubjective methods such as in-depth interviews are needed. When understanding the aims, customs and general reasons for actions of the involved practices is achieved, the next step is to explore what kinds of uses of the technology will be likely.

7.5 Step 4: Future Use of the Technology

As explained in Sect. 6.2, people have some implicit and explicit ideas about the practices of the eventual user, but these ideas do not necessarily have to comply with the eventual use. Therefore, step 4 enquires for which uses people will accept the new technology, based on the knowledge of their reasons for action described in step 3. Based on the preliminary information of the first few steps, several scenarios are imaginable concerning the use of ADAS technologies. Below, one example of such a scenario is presented.

7.5.1 Participating in Traffic

Given the reasons for actions of driving, what kinds of uses of ADAS technologies seem to be likely? The most important reasons for driving are freedom and fun: people find it important to be able to go wherever they want and to be anonymous in their cars. Future ADAS technologies will be highly autonomous in their actions and will both increase and decrease the desired fun and freedom. ADAS technologies increase fun and freedom in several ways: first, since the technologies take over some functions, they support elderly people (Davidse 2007) and less-experienced drivers. Secondly, with future ADAS technologies, people will have more freedom to focus on other primary and secondary tasks. As argued in 7.4, people also appreciate the time and space on their own in the car. Many commuters shave, eat, call and read in their cars, especially when they are stuck in a traffic jam. For them, ADAS technologies might be liberating in the sense that they can perform those secondary tasks more safely. Third, since it makes driving less tiresome, it might enable people to drive for larger distances or longer times. This will also increase the driver's freedom.

But ADAS technologies do not just increase fun and freedom: they also limit the freedom of drivers: especially the systems that take over the control of your car. When some actions are prevented and others are stimulated or enforced, freedom is limited and the fun might be reduced for those who consider driving to be a sporting activity. For them ADAS technologies might not be attractive.

Due to this mediation of freedom, there might be a difference in adapting the technology. A recent study concludes that although young people (the "early adaptors" of high-technologies) were initially more enthusiastic about ACC than older groups, after some test drives, the attitudes altered and the opposite became true (Davidse 2007). Automatic gearboxes aim to increase driving comfort, but hand-operated gearboxes are still the norm in Europe since many people feel that without switching gears by themselves driving becomes too boring. Cars with automatic gearboxes are mainly ordered by elderly people, who value the idea that without the need to switch gears, they have more time on their hands for other tasks.

These understandings may lead to different kinds of adaptations for different groups of drivers. It might be the case that some groups—such as elderly drivers and commuters—will gladly adopt the technologies, while other drivers may be more reserved. However, a differentiated adaptation of the technology may pose some problems. As explained in the introduction, it is hard to anticipate the behavior of another driver if you do not know whether he or she makes use of ADAS technologies like ACC (Hoetink 2003a). Since this is undesirable, the installation and use of ADAS technologies may be either forbidden or compulsory. Given the enthusiasm of governments, the latter is the more likely alternative (see also further on in this section).

Other road users such as pedestrians, cyclists and mopeds are often seen as passive victims, and so it seems odd to address the question of how they might make use of ADAS technologies. But when thinking this through, our imagination may draw attention to the possibility that pedestrians may also alter their behavior when advanced ADAS technologies become a standard feature of cars. If cars stop automatically, pedestrians may become less careful when crossing the road. A widespread introduction of ADAS technologies may also empower other road users. Systems that brake or slow down for the safety of the pedestrians and cyclists may create an emancipating effect. Beside, it is not unthinkable that NGO's supporting the interests of the vulnerable road users will ask governments to promote these systems.

7.5.2 Training for Traffic Participation

How will ADAS technologies be used in training people? The initial reason of training people was to make sure that they can operate all devices such as the pedals, the switches, and the gearbox. Similarly, when the training practice adopts ADAS technologies, it will be to learn how to program and use them. On average, people have to drive 400 km with ACC to learn how to use it safely. Nowadays, driving with ACC is not yet part of training new drivers (SWOV 2008b), but given the rapid developments it is to be expected that these technologies will also soon be used in the training practice.

Another use that is imaginable, though not yet studied, is that ADAS technologies might support learning. If ADAS technologies become simpler to use and develop a more integrated system, they may also be used to ease driving lessons. The more tasks the technologies fulfill, the more free time new drivers have to focus on learning other basic tasks. Used in this way, ADAS technologies can be used to support training by enabling new drivers to focus on a limited amount of tasks, making the first lessons less stressful.

7.5.3 Regulating Traffic

The main reason for action of government is to ensure road safety, to decrease fuel consumption and to reduce congestion. So, their regulation of use will be largely driven by the question of whether ADAS technologies will indeed increase road safety. As argued, the early versions of ADAS technologies suffered many disadvantages. Multiple simulation studies conclude that the traffic flow will only improve if a substantial amount of drivers makes use of ACC (>20%), (1) the speed is not too high or too low, (2) if the traffic is not too heavy, and (3) if the headway time is programmed at 1 seconds—which is unsafe. Furthermore, the claims that ACC reduces fuel consumption and increases car safety could not be sustained. But suppose the European Commission is right and that future, improved ADAS technologies will increase traffic flow, reduce fuel consumption, and improve safety; what might a government based scenario look like?

Here, the policy on seatbelts in cars may serve as an analogy. The first seatbelts were installed in the 1950s. Twenty years later, they were common and their contribution to safety had been proven in practice. Therefore, the government made it compulsory to install and use them (respectively in 1975 and 1976). Now, the same development might become reality for ADAS technologies like ACC. If the technology is improved and becomes generally adopted, it might become compulsory to install and use it. If only a part of the population uses the ADAS technologies, and behavior becomes hard to predict (Hoetink 2003a) and if vulnerable road users ask for ADAS technologies, then the government will be more likely to take such measures.

Regardless of whether these systems will become compulsory for non-professional drivers, it is imaginable that the government will stimulate professional use of these technologies (they already do). As we have seen in step 3, an important trend is to improve the safety and comfort of professional driving. If ADAS will indeed support a safe working environment for professional drivers, a technology push from the government is not unlikely.

Furthermore, as mentioned, the practice of regulation is not limited to traffic, and the question may be on which other uses the government may find for ADAS technologies. ADAS technologies are all information based, and some of them communicate with the infrastructure (like ISA) or satellites (like Dynamic Traffic Management). This information may be used for other purposes as well. As argued in step 3, the Dutch highway patrol also checks whether drivers have tax debts or use company cars for private purposes without paying the relevant income tax. The information obtained by the infrastructure of satellites may be used for these aims as well. Of course, this raises all kinds of privacy issues.

So far, this case study provided some examples of imaginable uses of ADAS technologies. Together these uses may result in the following mediation of reasons for action. First, different kinds of non-professional drivers might adopt the technologies in different manners, which may decrease the predictability of other people's behavior on the road. Consequently, partial adaptation may decline the road safety. Secondly, ADAS technologies protect more vulnerable road users, and this group may request that the government adopts a stimulating policy. Furthermore, the role of ADAS technologies in training new drivers might increase. If so, the government may make the installation and use of certain ADAS technologies compulsory. Now, the question is how this will mediate the fun and freedom of driving, the behavior of people on the roads, the safety of participating in traffic, and privacy issues.

7.6 Step 5: Mediation of Reasons for Action

Step 3 has shed some light on the reasons for actions in the affected practices, and step 4 focused on developing insights on how a technology might be used. This enables us to draw scenarios on how the technologies may mediate the reasons for action. In other words, when STS tools (like in-depth interviews, dialogue workshops and role-plays) and imaginative philosophy have illustrated what uses might be adopted, step 5 aims to obtain some ideas on how the adopted use of a technology may alter human behavior by questioning how the technology will mediate the reasons of actions.

Thus, the last question is how the likely uses of ADAS technologies might mediate the reasons of actions of the involved practices. Three different kinds of mediation of reasons for actions are the mediation of options, perceptions and moral beliefs. This section is structured according to the distinct groups of involved practices, using the idea that technologies mediate these three kinds of reasons for action as a background theory.

As explained in the previous chapter, to understand how a technology may alter a reason for action, one might interview people who have experience with prototypes. Correspondingly, to find out why people respond in a certain way to new technologies, behavioral field studies on road safety are sometimes complemented with intersubjective research methods. Most of these intersubjective studies consists of interviews which ask for instance questions like (Oei 2003) "How do you think your behavior was influenced by the technology (overtaking, speed, attention)?", and "do you appreciate the technology for its comfort, safety and financial costs?" Such methods can provide information on the way in which technologies may mediate the driver's reasons for actions, though (as argued before) imaginative philosophy is essential for taking this step. The next section briefly introduces how taking such a step may work.

7.6.1 Participating in Traffic

For driving, several kinds of mediation of reasons for actions are imaginable. Let me, discuss three examples of possible forms of mediation that are connected to the three reasons of actions: (1) the mediation of what it means to be a driver as a form of mediation of perception, (2) the mediation of change in road use as an example of mediation of options for actions, and (3) the mediation of responsibility as an illustration of how moral beliefs are mediated.

One question is, if the intelligent car of the future becomes the norm, will the difference between being a driver or a passenger become smaller? On the BBC news of January 3, 2009, a reporter tested an automatic speed-limiting device. He drove 60 mph while passing a 40 mph sign, and the device took over and slowed the car down. His spontaneous response was: "O... o... that is really quite strange... It is almost like if someone else is driving." A similar situation occurred in the television show Top Gear, when Jeremy Clarkson drove an Audi Q7 that took a roundabout almost fully automated. On the one hand, Clarkson was really impressed: when he approached the roundabout, the system switched gears, and reduced speed from 60 mph to a full stop. When he steered, the car gently took the roundabout, and after leaving, it accelerated to the initial speed again. Clarkson only needed to steer: he did not have to use the pedals, nor the gears. But on the other hand, Clarkson also felt superfluous and asked himself "What am I here for?"

This is a mediation of microperception, which will also alter macroperceptions. The more advanced and integrated ADAS technologies become the more the driving experience will be likened to taking a train, especially when the full potential of Car2Car Communication is realized. Consequently, people will start to perceive themselves more as passengers than as drivers.

For some, this will increase the joy of driving: they can use their attention for other tasks or for communicating with other people. But for people who like to drive (as sport or recreation) this raises questions about the added value of the technologies. Furthermore, many people are proud of their driving style and see their cars as a status symbol. People also lend status to technological tools like navigation (Tabula Rasa 2004). However, it is questionable whether the young-and-dynamic-image the industry aims for is what these users will find in a future car that diminishes the role of the driver. The question arises: how does the aim to improve one's status by being high-tech interfere with the aim to improve comfort and safety? Will people use ADAS technologies as image statements or will using ADAS technologies provide you with an image of being dull and non-capable? Many car commercials stress that having the latest high-tech novelties is good for your image. Having the newest technologies is associated with sporty driving styles and with being successful and self-aware. But can you be sporty or express your independence with a car equipped with so many technologies that it drives itself and only needs "passengers" to tell it were to go?

The diminishing difference between passengers and drivers is also a mediation of options for action. It is imaginable that some people will drive more, while others will make less use of the car. If driving becomes more like taking a train or a taxi, the recreational use of cars is likely to decrease. However, many people travel by train because it is less tiring and because in the train they have more time on their hands for other tasks such as reading, listening to the radio, eating, and so on. But trains are limited in the sense that they run from station to station. Cars provide you with the option to drive from door-to-door. If ADAS technologies (including Car2Car Communication) make cars of the future more like a private train combining the advantages of both contemporary options for commuting in one mode of transport, road use might also increase.

If more people drive instead of take the train, fuel use and congestions may increase instead of decrease. Research has shown that fuel consumption is likely to reduce since ACC generates a more constant gas flow than can be reached with manual drive (Hoetink 2003a). However, this positive effect can be diminished if more people commute by car.

Finally, when the car of the future dissolves the differences between being a driver and being a passenger, drivers may also start to behave like passengers and might become less willing to accept certain responsibilities. In other words, it might mediate our moral beliefs. How will the attitudes of motorists alter toward pedestrians and cyclists, and how will the vulnerable road users behave toward intelligent cars? As some road safety studies have showed, when ACC is installed in their cars, drivers show more asocial behavior toward other, more vulnerable road users (Hoetink 2003a). Drivers wanted to avoid having to reinstall the speed (the early model required reinstallment following every use of the brakes), but also the automatic deceleration of the car irritates some drivers.

However, pedestrians and cyclists may exploit the feature that cars will automatically stop for them. They can safely take right of way, and they can behave more assertive. Thus, on the one hand, drivers may become less social toward this group. On the other hand, vulnerable road users may become more emancipated and may show more risk-taking behavior on the roads (for the cars will slow down anyway). Consequently, power relations alter, and responsibility issues may shift with them. As argued, technologies not only take away our responsibilities, they also generate new responsibilities. Other road users (and especially the vulnerable road users) need to be able to trust and anticipate your driving, even when sitting "behind the wheel" becomes more like programming a computer. Thus, drivers will have a moral duty to make sure that the technologies are properly installed and up-to-date. These new responsibilities will also require other skills and other infrastructures (for instance, the possibility to connect to the internet).

7.6.2 Training for Traffic Participation

For training for traffic participation, two examples of mediation of reasons for actions can be provided: the mediation of option in relation to examination requirements, and the mediation of perceiving adolescents. Both of these forms of mediation can be pointed out by asking: if an increasing number of ADAS technologies are installed, what does this mean for learning driving skills by future traffic participants? In the Netherlands, the legal age to start driving lessons is 18 years. The exam consists of a theoretical and two practical parts. Both the theoretical lessons (not compulsory) and the driving lessons (on average 35–40 h) are provided by governmentally certified driving instructors. The exams are set by the Centraal Bureau Rijvaardigheidsbewijzen (CBR). Now the question is, if the scenario of step 4 becomes reality, will it still be necessary to subject people to so many driving lessons and exams?

Since driving will become more about programming information systems and less about internalizing new skills, it might become an option to loosen regulations regarding examination. If the car of the future enforces responsible driving, why still stick to the rule that you should be 18 years old before starting driving lessons? Why not allow 16 year olds to drive, if the driver's responsibility is restricted to programming the information systems? This would generate many more options for actions for this group.

Such a decision would have a tremendous social impact, as this group will become more mobile and more independent. They will be able to choose schools and jobs that are difficult to reach with public transport. But it might also mediate how we see teenagers: in other words, it might mediate our perceptions of what adolescents are capable of and responsible for. In the US, the minimum age for receiving a driver's license is 16 years. The 16 year olds are seen as more mature. This allows them to have jobs further from home, to go independently to sports clubs and parties, and to drive to school. In many European countries, where the minimum age for driving is 18 years, such independence of youngsters is unknown, and adolescents are treated less maturely. If new technologies lead to the permission for younger people to drive, it is not inconceivable that we will also treat them more fledged in other respects as well, assigning responsibilities to them such as taking their younger siblings to school or sport.

7.6.3 Regulating Traffic

The most eye-catching example of mediation of moral beliefs may be found in regulating traffic. If the government stimulates a widespread introduction of ADAS technologies, they also have to deal with multiple responsibility questions, see also (Waelbers 2009c).

First, if ADAS technologies become compulsory, then what will we regard as responsible behavior and will drivers accept responsibility if technologies perform the tasks? If technologies take over human responsibilities, our moral judgments of traffic situations need adaptation too. ADAS technologies inform drivers and intervene in the driving. What if they misinform or fail to intervene adequately? Who then bears the guilt: the driver (for he or she should pay attention, regardless of how many technologies support him or her) or the manufacturer (who should deliver infallible products)?

Next, if cars enforce responsible driving behavior, the task of the highway patrol is likely to change: checking vehicles will become more effective than controlling the behavior of drivers. For instance, existing prototypes of governors (speed limiting devices) can communicate with satellites and prevent motorists from speeding. If these devices become standard, no one will be able to speed. In such a case, hiding mobile brigades with cameras in the bushes along the highways is no longer of any use. It would be much more effective to organize large-scale screenings of cars: are the devices working properly? And did anyone try to sabotage his or her governor? Here, an analogy with the tachograph can be made: since they became compulsory in buses and trucks, the police focus strongly on checking the data that are printed on the round cards.

But responsibility is not just about liability and enforcing regulations; the government should also reflect on attitude, which can alter when important tasks are delegated to technologies. One imaginable change is that people may regard something as acceptable if the technologies allow it. Compare this with speed limits that were initially only meant to regulate safety and encourage people not to drive too fast. Nowadays, motorists are irritated if others drive 40 mph on a road that allows 50 mph, and the police also issue fines for "slow driving". In the same way, technologies that identify the minimal safe headway time may become the norm as the correct headway time. We need to address questions like: will drinking and driving become acceptable again if the Driver Drowsiness Monitoring and Warning system proves no decrease of driving skills?

We may conclude that the enthusiastic response of the European Commission needs to be reconsidered if we want to work toward a responsible introduction of ADAS technologies. We might gain more safety, comfort and permeability, but we might also lose some things. The safer the cars are, the less safe (daring or unfocussed) people may behave. We might also lose the fun and the feeling of freedom that driving brings to many people, though drivers might have more time for other primary and secondary tasks. The vulnerable road users may become more emancipated, but we are also at risk of assigning drivers' responsibilities to technologies.

Table 7.2 Some examples of the mediations of reasons of actions that might be induced by the intelligent car of the future

Perceptions	Options for actions	Moral beliefs
Microperceptions	*Former options disappear*	*Values*
• Who is driving?	• Will it be possible to have a sporting driving style?	• Will it be responsible in the future to drive without supporting intelligent technologies?
Macroperceptions	• Will some technologies become obligatory?	• How will the meaning of freedom change?
• What will the difference between being a driver and a passenger be?	*New options arise*	*Norms*
• What does it mean to be a 16 year old?	• What new options and lifestyles will become feasible for younger, older and disabled people?	• What new duties will arise for maintaining and updating the IT based technologies?
• Is being high-tech still about being cool and dynamic?	• Will the government alter the traffic regulations and the way to enforce them?	• What new duties will we assign to teenagers?
	• For what other uses (like taxation) might the government employ the information the technologies generate?	• Will driving and drinking become acceptable?
	Existing options alter	*Virtues*
	• Will we be able to call, eat and read safely behind the wheel?	• Do drivers take responsibility for accidents?
	• Can pedestrians become more emancipated?	• How will the attitudes of drivers toward other road users change?

Drivers may become less committed to the vulnerable road users. New standards of what we consider good road safety behavior may be set. These standards might be better in some respects, but they will be more machine based (how quickly can the devices respond) and less human based (how quickly does the average driver respond?). Young people may gain freedom if driving exams become less harsh, and if they are allowed to drive before they are 18 years old. But what does this mean for their upbringing and our society?

The five-step approach offers a way to structure the enquiry, which raises and discusses multiple questions on how a technology might be used and what this implies for our reasons for actions (see also Table 7.2). The behavioral studies and interviews that are held can answer some of these questions, but much is still to be researched. Many questions also ask for more intersubjective input. But nonetheless, based on the limited material available, the five-step approach already produced both obvious and unexpected insights in the possible future social role of ADAS technologies. It also nuances the optimistic outlook of the European Commission by pushing forward multiple questions to deal with the responsible introduction of those technologies. Now the question remains, how to evaluate these findings?

7.7 Evaluation

How to evaluate the results of the five-step approach during for instance a consensus conference or dialogue workshop? Is the optimism of the European Commission tenable? The possible mediations of actions suggest that we should make a more nuanced evaluation. But how can we do this?

For evaluating the insights generated by discussing the five steps, Sect. 6.4 formulated some starting points. These starting points were based on the premise that the aim of developing and employing technologies should be to promote human and environmental flourishing. However, the starting points are not moral rules one has to obey: they are based on different types of needs to lead a flourishing life. One should have strong arguments for developing or employing a technology that prevents the realization of one of those elements. However, interpreting and weighing the distinct elements should be done by people from the involved practices as there is not one universal way of leading a good life. How can they do this? How can the presented toolbox provide some starting points for the discussion? Based on the premise of human and environmental flourishing, pragmatic and substantial criteria were formulated. This last section, presents how the pragmatic and substantial criteria can be used to generate questions for an intersubjective evaluation of ADAS technologies.

7.7.1 Pragmatic Criteria

Section 6.4 formulated two pragmatic criteria for the evaluation. First, the problems the technology creates should not be bigger than the problems the technology intended to solve. Secondly, a scenario is more desirable if it leaves more choices open in the future. The following subsections discuss both points.

The New Problems Should Not Be Bigger Than the Problems That the Technology Tried to Solve The first pragmatic evaluation criterion is based on the idea that possible undesirable social roles of technologies should be either restorable or outweighed by the desired social role. This criterion includes what Edward Tenner called "revenge effects". To see whether this is the case, the answers to steps 4 and 5 can be compared with the aims of the technology (step 1).

Step 1 worked out four explicit aims: increasing safety and comfort, and decreasing congestion and fuel consumption. Some preliminary empirical studies have shown that ADAS technologies such as ACC systems improve road safety in some situations, while reducing safety in others (Hoetink 2003a). In this sense, it is vulnerable to the "revenge effects" principle. These studies involved early versions of limited kinds of ADAS technologies, and so their results may not be directly translated to evaluate the car of the future. But still, the results of these studies show that it is not unlikely that ADAS technologies will have revenge effects on road safety. A responsible introduction of the technology should address this problem by taking into account either how such a revenge effect might be prevented, or by putting the question forward of whether the expected benefits for road safety outweigh the possible negative outcomes.

Comparable arguments can be made when it comes to comfort, congestion and fuel consumption. Step 3 explained that comfortable driving also has to do with optimizing arousal level: people should not be overloaded with tasks, but they should also not become bored because the technologies take over all tasks. Congestion may be decreased if multiple ADAS technologies (including Car2Car Communication) are adopted, since the traffic flow would be more efficient. But suppose step 5 is correct and more people might take a car instead of a train, since the new car of the future will combine the best of both worlds. If so, congestion might increase and so might fuel consumption.

Note, that the aim of this step is not to say that all technologies will create bigger problems than they are designed to solve. It might very well be the case that further study will show that some of the revenge effects are unlikely. Nevertheless, raising questions about all possible revenge effects in the evaluation is important when taking responsibility for the social role of new technologies.

A Scenario Is More Desirable If It Leaves More Choices Open in the Future The second pragmatic criterion has to do with choices. Because it is fair to allow future people to make their own choices, and because later in time more will be known about the possible social roles of a technology, it is generally better to strive for an option that leaves more choices open for later.

The example in step 5 provided three reasons for making ADAS technologies obligatory: (1) the diversity in adapting ADAS technologies made it hard to anticipate the behavior of other drivers, (2) the vulnerable road users felt well protected by the new technologies and requested for such a measure, and; (3) it improved the working conditions for professional drivers. An opposite development is also conceivable: since the diversity in adapting ADAS technologies made it hard to anticipate the behavior of drivers, the government decides to forbid these technologies (although, given the current enthusiasm of governments, this might not be a likely scenario). Both such options do not leave much choice for the future. If people find out that the choices made have undesirable social impacts, it is hard to alter course. When a technology is forbidden for a while, the industry will stop developing and producing it: knowledge and infrastructure will be lost. But when a technology is widely adopted, it becomes difficult to replace it because of the embeddedness in the behavior and infrastructure (remember the example of the inefficient QWERTY keyboard discussed in Sect. 2.1).

Thus, choices that leave more options for actions open seems to be more desirable. Such a course is not difficult to imagine: nowadays some people prefer automatic gearboxes, while others like to switch gears themselves. An ADAS scenario that leaves the choices open to the individual driver can become reality. However, as argued, not knowing that someone uses ADAS technologies makes it harder for other traffic participants to predict his or her actions as a driver. So the question is, which scenario do we find to be the most desirable?

7.7.2 Substantive Criteria

For deciding which option we desire the most, we need to evaluate the distinct elements of the possible future developments. Section 6.4 argued that MacIntyre's four elements of having and living a good life are useful starting points for discussing whether a social role of a technology can be called good. These four elements are:

- Biological vulnerabilities
- Skills
- Social roles
- Virtues

The next subsections illustrate how these elements can be used to kick off the evaluation of the car of the future.

Biological Vulnerabilities The first element of a "good life" has to do with the idea that we are biologically vulnerable and that we have to fulfill our bodily needs. Technologies should be evaluated with regard to the protection of our biological vulnerabilities. Here we arrive again at the discussion on the question of whether ADAS technologies will increase road safety.

But, vulnerabilities are not a-historical: technologies mediate our needs. In the car industry, the trend of constantly increasing safety norms is clearly recognizable. Since the 1970s, the number accidents on the Dutch roads has decreased every year (see Fig. 7.1). In 2008, we had about the same number of casualties as in 1950, while the number of vehicles increased from 0.14 million to just over 10 million. Thus, we may conclude that participating in traffic has never been as safe as it is nowadays. Initially, both the number of cars and the number of casualties increased. But since the 1970s, numerous improvements in car technologies (such as seatbelts), road design, training of new drivers, and anti-alcohol campaigns stimulated a substantial reduction in fatal accidents. In 2008, 750 people died due to traffic accidents, and the government is still working to lower this number.

The question is to what extent ADAS technologies may contribute to these aims, since ADAS technologies are also unable to prevent all accidents (like those caused by oil on the roads, failing brakes and suicides) and since there are many possible revenge effects. And then the question arises: do the benefits outweigh the disadvantages? In other words, does the decrease of accidents outweigh the doubts that arise when discussing the other three elements of a flourishing life?

Skills The second element of a flourishing life is the possession of intrinsically valuable skills or techniques. The question is whether a technology indeed facilitate skills and techniques in a desirable manner. If the difference between being a driver and a passenger decreases, a loss of skills and a loss of situation awareness may be the result (Davidse 2007). If people do not have to perform certain tasks regularly, they will become less skilled in those tasks. When people have been using automatic gearboxes for several years, they have to get used to switching gears manually again. Now, if you are used to an LDW that increases the resistance of the wheel

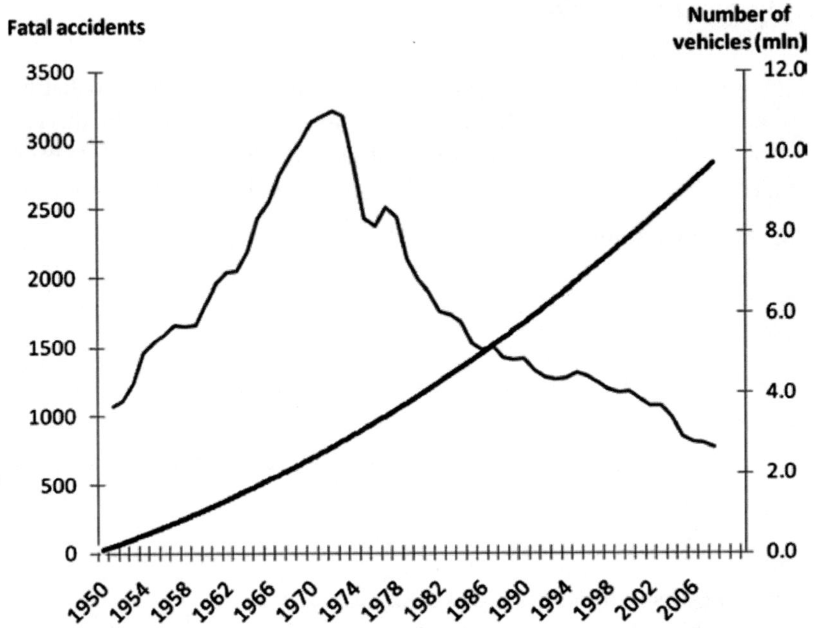

Fig. 7.1 Increase of number of vehicles plotted against the decline of fatal accidents on the Dutch roads since the 1970s (Source: Statistics Netherlands [Centraal Bureau voor Statistiek, CBS], 2009)

when shifting a little to the left or right, you will find it harder to steer a normal car. A similar argument can be made for situation awareness: estimating distances and knowing when to break will become less second nature if you are used to a technology that does it for you. Taking all ADAS technologies together, the impact on skills is substantial.

Thus, if the car of the future becomes reality, we may lose some valuable skills. But is the value of these skills intrinsic or instrumental? One could argue that we mainly need those skills for road safety. According to this view, if technologies can ensure at least the same level of safety, we only lose instrumental skills. But there are also people that see driving as a sporting or recreational activity that provides them with fun and the feeling of freedom. For them, executing and perfecting skills is intrinsically valuable. How to deal with this diversity?

When introducing ADAS technologies, this study mentioned that many of these technologies have different variants: some are "open" in the sense that they just warn the driver with sounds or lights. The driver can merely ignore these signals. Next, there are "semi-open" systems that increase for instance the resistance of the steering wheel or gas pedal. Finally, "closed" systems enforce certain actions like braking or they prevent the driver from steering in certain directions. To enable people to enjoy practicing driving skills, we might argue for the choice of the more "open" variants of ADAS technologies. However, again we need to weigh the value of driving as a fun or sporting activity against the increased unpredictability of behavior for other drivers if they do not know which technologies are intervening.

Social Roles Technologies can mediate the value of the social roles that are pursued in society. Following the Luddist, we can argue that the intelligent car of the future may make professional driving a considerably less interesting activity. But as mentioned, new technologies can also emancipate people, providing them with the opportunity to fulfill valuable social roles. Such arguments may outweigh the Luddist arguments. As argued already in step 4, ADAS technologies may support elderly drivers. Existing research suggests that those technologies may enable older drivers to participate better and for longer in traffic (Davidse 2007). For many people, being able to drive means being able to participate in social life, and for them, the advantages of the car of the future would be highly valuable. A similar argument might hold for people with certain disabilities who are now not able to participate safely in traffic. The car of the future might enable them to become less dependent on other people for transport.

But as step 5 argued, ADAS technologies might not only support the participation of older drivers, it might also support training young drivers. Step 5 mentioned a possible effect on the legal age for driving: why wait till people are 18 years old, if the intelligent car of the future ensures responsible driving anyway? This will increase their mobility and independence, and consequently it might also change our perception of what teenagers can and should do. On the other hand, it also reduces the abilities of parents and teachers to control the actions of their teenagers. During the evaluation we need decide how to value these possible positive social impacts, and how to weigh them against the negative impacts.

Morality MacIntyre's last element of a flourishing life is that you should be able to become a good and virtuous person. The social role of a technology can be evaluated for its potential to support people's abilities to be and do good. The previous chapter frequently argued that at least a limited amount of freedom is needed to live and act according to one's own moral beliefs. This is a central characteristic of what it means to be a human actor. The question is, on how to evaluate the mediation of our moral beliefs by ADAS technologies?

First, there is the question of how we evaluate the shifts in responsibility if cars of the future are taking autonomous decisions. Is it fair to hold the drivers responsible? Probably not. And thus the question is how are we going to deal with a situation in which humans cannot be held responsible for their driving. Though here we should not make the mistake to assume that technologies only limit our morality, they also generate new moral norms and responsibilities. Although the new moral norms will be more technology based, they still require drivers to take the responsibility to install and maintain ADAS technologies. They have to ensure that they are safe, and they need to respond adequately on the ADAS feedback, which both generates new responsibilities. The question is how we evaluate this shift.

Second, there is the question of whether we can be virtuous in the traffic if technologies make the cars drives more or less autonomously. Being virtuous in traffic involves things such as giving way to someone who (according to the regulations) actually has to wait for you, and being patient if someone makes a mistake or needs to unload some goods. As argued, ADAS technologies may lead to more asocial and impatient behavior. Preliminary studies show that installing relatively limited ACC systems already lessens the orientation toward other road users, motorists as well as more vulnerable road users (Hoetink 2003a). If a wider variety of high-tech ADAS technologies become standardized, drivers may become more and more focused on the feedback given by the car and may become less oriented on other people. Evaluating the social role of technologies implies questioning how we value our moral agency in traffic.

Last, the questions on the aims and means of regulating traffic need to be addressed. The car of the future will likely require new ways of control. An important question for the evaluation is, do we want to live in a world in which satellites and road infrastructure is employed to steer your car *and* to register your whereabouts? For what other goals do we want such information to be available?

Summarizing, if the European Commission is interested in a responsible introduction of the car of the future, they have to consider not just the promises of these technologies. Raising questions about all possible revenge effects in the evaluation is important when taking responsibility for the social role of new technologies. Multiple complex questions need attention, such as: To what extent will the technologies really increase safety and comfort, and decrease congestion and fuel consumption? And, does this outweigh the possible loss of skills and the loss of freedom and fun many people experience when driving? What kinds of ADAS technologies do we want? Do we prefer "open" technologies that allow people to drive themselves and enable us to have more options for actions in the future? Or do we consider the predictability for other road users as more important? And, finally, how do we evaluate the possible shifts in our moral agency, morality and social roles? Taking responsibility implies taking these kinds of questions into consideration, and does not allow for an uncritical promotion of new technologies.

Chapter 8
Will We Accept Responsibility?

This study began by explaining how philosophers such as Sunstein and Thaler argue that if technologies are mediating our actions (whether we design them to do so or not), we should take responsibility to make sure that this mediation is for the better. However, this suggestion suffers from two conflicting findings. As Latour made clear, the human-technology interaction is a complex black box of which the outcomes are often hard to predict. We cannot simply *design* the mediating role. Furthermore, the technological change of our actions prevents us from being autonomous, rational agents, who are free to determine our own course of actions. These conflicting notions led to the main question of this study: Can practitioners—such as scientists, engineers, commissions, R&D institutes, etc.—take responsibility for the social roles of technologies if these technologies mediate our actions (and therefore limit our autonomy) in ways that are hard to predict?

This overarching question was broken down into three questions: (a) Can we still be responsible when technologies co-shape our actions? (b) How can we best understand this social role? and (c) How can we work to take responsibility for the social role of future technologies in practice? Following subsections summarize conclusions regarding these three questions. A concluding note explores whether people will accept this responsibility for the social role of technologies or whether other problems still need to be resolved.

8.1 Responsible Beings

The first conclusion that can be drawn is that although we are not free, autonomous and rational beings, but are still capable of taking responsibility. Latour showed that technologies co-shape human action (see Chaps. 2 and 3). Furthermore, we are embedded in existing traditions that form our views, and our physical constitution provides us with needs such as food, shelter and medical care. MacIntyre stresses that since we need these essential goods, we need others to survive and flourish

K. Waelbers, *Doing Good with Technologies: Taking Responsibility for the Social Role of Emerging Technologies*, Philosophy of Engineering and Technology 4,
DOI 10.1007/978-94-007-1640-7_8, © Springer Science+Business Media B.V. 2011

(for more details, see Chap. 4). And thus we are biologically vulnerable and socially interdependent. Both Latour and MacIntyre have found—in their own distinct manners—that the autonomous, rational human agent is a ideal which we will never fully realize. We are biologically vulnerable, socially interdependent and technologically mediated. So how can we ever be responsible for anything?

The autonomous subject presents an important ideal to strive for: many people value rationality and autonomy. To increase one's autonomy (even to a limited extent), one has to understand the boundaries of this autonomy: if you know the borders you are fighting, you have a better chance for pushing them back. MacIntyre's distinction between reasons (motives) and practical reasoning (reflection) is helpful for achieving a little autonomy, without assuming an atomistic, autonomous subject. He explains how our biological condition and social surroundings provide us with reasons for actions: we are hungry, and so we eat. We are born into certain traditions, and so we hold certain beliefs. But this does not imply that we should eat every time we feel like eating, or that we should believe everything other people say. "Being responsible" means that we are using our capacity of practical reasoning to reflect on the given reasons. Practical reasoning is not an individual, transcendent, rationalistic endeavor. Nor is it the same as striving after one's own interests, which the power-oriented ANT stresses. It is a social process that implies communication with others and takes given reasons seriously.

Practical reasoning often takes place in the practices. Practices are socially established cooperative human activities through which internal goods, such as virtues, are realized. To be a successful member of a community of practitioners, one has to adopt most norms embedded in the relevant practices, including most moral norms. By using the vocabulary of practices instead of networks and stakeholders, more room is provided for discussing a wide palette of reasons. The concept of practices allows for the inclusion of desires, intentions, and moral reasons, which is crucial for exercising moral agency (and so for taking moral responsibility).

So, even though our reasons are biologically, socially and technologically shaped, we can take responsibility in the sense that we are able to enquire whether the reasons for our actions are desirable enough to pursue those actions. Of course, we do not have to reflect constantly on everything we do. Many actions are well established habits that are absolutely fine. But some actions do need reflection, such as those that involve new technologies or new employments of technologies.

8.2 Technological Mediation

A second conclusion of this study is that we should take the mediating role of technologies seriously, even though ANT is wrong in claiming technologies are agents. In ANT, technologies are called agents because their influence on people and other technological developments can be traced back in history. We can, retrospectively, understand what the social role of the technology was. This role is describable in socio-anthropological language similar that we use for describing the role of humans or organizations. However, technologies lack the capacities needed for exerting agency, since they have no intentions, desires or means of practical reasoning. Technologies are not actors, but they are social factors. In order to take responsibility for these social factors, it is important to understand how such factors mediate our actions.

Chapter 5 argued that existing mediation theories are often too simple since they only recognize a limited range of technological mediation of actions or ideas. However, by recognizing that technological mediation starts with mediating reasons for actions, we can obtain a fuller appreciation of how this might work and what the different types of mediation might be. Whether we are conscious or unconscious of those reasons, the social role of technologies needs to be explained by their effect on our reasons for action. The step we have to take from here is to see what kinds of reasons for actions we have, and ask ourselves whether these can be technologically mediated.

Since there are three kinds of reasons for actions—is, can and ought—there are three mechanisms of technological mediation of reasons for actions. First, the post-phenomenological analyses of Ihde make clear that our perceptions (our factual ideas) are mediated by new technologies. Chapter 7 explained for instance that the intelligent car of the future can alter our microperception of what is the case with regard to whether teenagers are able to drive safely and so our macroperception may change on what tasks and freedom we may allow them.

Secondly, technologies alter our options for action: previous options disappear (like the option of not having e-mail) and new options arise (like the option to determine the gender of your future child). Also the discussion of the car of the future provided some examples of how the option for actions may be mediated. Since a broad introduction of a palette of ADAS technologies may release drivers from many duties, they will have more time for secondary tasks such as eating, listening to music or even reading (see Chap. 7). If so, driving will become more like taking a train which takes you from door to door, and more people might take the car because of this new option.

Third, as explored in NEST-ethics, our moral beliefs are mediated by technologies. For example, the sexual revolution of the second half of the twentieth century was amplified by the invention of the contraceptivepill and the improvement of condoms (Swierstra 2010). Similarly, Chap. 7 showed that the car of the future, may mediate our morality in the sense that it alters our virtues and beliefs on responsibility issues related to driving and raising adolescents.

Thus technologies mediate our reasons for actions, but that does not imply that technologies are agents. Nor does it imply that we lose our capacity of practical reasoning—that is, as long as we are not discussing mind-altering technologies such as deep brain stimulation, psychopharmaceuticals and brain extensions. In this sense, the reformulation of mediation theory is less controversial than ANT. But, recognizing that technologies mediate our actions by altering our reasons also has a more radical consequence: our moral beliefs can no longer be described as fixed pillars that can be rationally understood. We cannot use pre-existing moral rules to guide our employment of technologies. On the contrary, our moral beliefs are technologically malleable. How can we take responsibility for the technological mediation of our reasons for actions?

8.3 Taking Responsibility in Practice

This leads to the third conclusion: we can only take responsibility if we have the tools that support imagining the possible technological mediation of our reasons. So far, philosophy has offered few tools to help people anticipate the possible social role of technologies. This study aims to contribute such tools that may facilitate practitioners to take responsibility in practice. To structure the reflection on the possible future social roles of technologies, a five-step approach was developed (see Chap. 6). This five-step approach is based on the idea that technological action in one practice has consequences for other current and future practices. The five steps thus need to address the relation between the aims and functionality of the development or management of technology and the social impacts other practices. To this aim, the five steps consist of the following questions:

1. What are the aims of the technology?
2. Which practices will be affected?
3. What are the common reasons for actions in those practices?
4. Given these reasons and given the existing technologies, what uses are likely?
5. How will these uses mediate the reasons for actions in the involved practices?

To answer the five questions, the toolbox introduced several types of research methods to support the needed imagination, such as imaginative philosophy, intersubjective research methods of STS, and empirical research methods of behavioral sciences. Chapter 7 illustrated how these five steps and the distinct research approaches can serve to qualify the optimistic view of the European Commission regarding the autonomous car of the future. It showed that none of these academic approaches can foresee what the future social role of the technology will be: there is no such thing as a crystal ball. Yet combined, they are able to produce interesting insights into the possible mechanisms of technological mediation. Based on these insights, we can sketch techno-social and techno-moral scenarios.

To evaluate these scenarios, Chap. 6 offered two starting points for discussing such a normative framework, both based on the presumption that technologies should commit to human and environmental flourishing. This normative framework combined pragmatic and substantial criteria. Pragmatic criteria teach (1) that the problems new technologies might create should not be bigger than the problems the technology was supposed to solve, (2) that possible substantial problems should be reversible, and (3) that a technology that leaves more options for actions open, is commonly more desirable. Next to these three pragmatic criteria, more substantial criteria can be found in MacIntyre's description of the four goods needed for the realization of a good life. Interpreting these four goods is susceptible to technological mediation and needs further specification in concrete contexts. Nonetheless, these four goods can be used as a starting point for discussion on the desirability of the possible social role of a technology.

8.4 Accepting Responsibility

Generally, ANT studies that focus on mediation do not raise the question of how to take responsibility for the future social role of technologies. This study works to show that moral theories can add some valuable insights to ANT. In doing so, the difficulties ANT poses for accepting responsibility are reduced to a manageable level. But also the opposite was aimed for: even though good life theories do recognize that moral beliefs are not a-historical, they rarely address technologies as social factors. It would enrich good life ethics if the insight on technological mediation were to be more widely adopted. This study developed some preliminary ideas of how this may be done.

Hopefully this study will prove relevant for both ANT and good life ethics, but above all, the aim is to make complex theoretical insights of practical use. By developing a practical toolbox, this philosophy of technology aims to stimulate practitioners involved in developing and managing new technologies to actually take responsibility for the social role of technologies. Taking this responsibility is difficult, but the tools can support this task. But does the availability of a toolbox imply that practitioners will accept this responsibility?

Many people are involved in the development, management and use of technologies: scientists, engineers, companies, governmental organizations, NGOs, commissioners, and users all work with and against each other to influence the outcomes of socio-technological changes. Consequently, it is difficult to foretell whose actions will lead to which outcomes, and the question of who should take which responsibility can become blurred. This problem is called the problem of the many hands (Thomson 1980), and the question is "will practitioners accept responsibility as long as this (unsolvable?) problem remains unaddressed?"

Although it is difficult to foresee the impacts of individual actors, the problem of many hands should not be turned into the "Huey, Dewey, and Louie excuse". When Donald Duck gets angry with his nephews and asks who shot the ball through the window, Huey points at Dewey, Dewey points at Louie, and Louie points at Huey, providing Donald Duck with an impossible puzzle. Also in relation to the social and environmental impacts of technology, people often point at others. Natural scientists often claim that they merely develop knowledge and that practical applications are not within the scope of their responsibility. Engineers and engineering companies pass on the responsibility by arguing that there is nothing wrong with their devices. In their eyes, the use (or misuse) of technologies is the cause of the problems. Users, in their turn, tend to claim ignorance and lack of power: how can an average consumer know what the impacts of technologies might be? Even with knowledge, what can individuals do? They often point out that the government should take responsibility. However, western governmental institutions often act slower than many technologies develop. Moreover, they are reluctant to interfere with both technological progress and the free market economy. In capitalist societies, governments often start by pointing out the responsibilities of companies and consumers.

What is the use of the five-step approach, if most groups immediately start pointing at each other when the question of responsibility is raised? First, it is important to keep in mind that the five-step toolbox is not about finding scapegoats. As the issue is not who to blame taking responsibility involve being prepared to enquire what the possible social role of the technology will be. So, the toolbox does not address the question of "who is liable for the undesired consequences we experience from technologies in our societies?" The request is to think about how we can do a better job for the sake of the future. It does not make sense to address only one actor: we need to support the STS aim to stimulate debates among a wide variety of actors.

Secondly, the current struggle with the negative social impacts of many technologies shows that it is prudent in the long-term to take responsibility for the social role of technologies. It is no use picking one person or one group of persons and assigning this form of responsibility to them. The term "we" occurs quite often in this study because taking responsibility only works if many actors work together in developing new technologies. Intersubjectivity is needed for exploring which course will lead to human and environmental flourishing, as it is undesirable that a small minority determines what a good life entails for all. So together, people should take up the task of enquiring into the social impacts of technologies.

In real practice, it may become clear that the five-step toolbox needs further adjustment, and of course, we will always be imperfect in imagining the future social roles of technologies. However, this is no reason to give up on taking responsibility. We have a better chance of realizing a desirable future if we make educated guesses, than if we simply implement new technologies or new uses of technologies without using our capacity of moral imagination at all. So let us take this responsibility seriously for the sake of a flourishing future.

Literature

Achterhuis, H. (1993). Het permanente gemoraliseer. In H. Achterhuis (Ed.), *Deugt de ethiek?* (pp. 102–116). Baarn: Gooi en Sticht.
Achterhuis, H. (1995). De moralisering van de apparaten. *Socialisme en Democratie, 52*(1), 3–12.
Achterhuis, H. (1998). *Erfenis van de Utopie*. Amsterdam: Ambo.
Akrich, M. (1992). The de-scription of technical objects. In W. Bijker & J. Law (Eds.), *Shaping technology, building society: Studies in sociotechnical change*. Cambridge: MIT Press.
Anders, G. (1980/1956). *Die Antiquiertheit des Menschen*. Munchen: C.H. Beck.
Anscombe, G. (1981/1958). *Modern moral philosophy* (The collected philosophical papers of G.E.M Anscombe, Vol. III). Oxford: Blackwell.
Applbaum, A. (1999). *Ethics for adversaries: The morality of roles in public and professional life*. Princeton: Princeton University Press.
AVV. (2001). *ISA Tilburg: Intelligente Snelheids Aanpassing in de praktijk getest*. Den Hague: AVV.
Barnes, D. (2006). *The great stink of Paris and the nineteenth-century struggle against filth and germs*. Baltimore: John Hopkins University Press.
Basalla, G. (1988). *The evolution of technology* (Cambridge History of Science series). Cambridge: Cambridge University Press.
Bentham, J. (1789/1996). Introduction to the principles of morals and legislation. In J. H. Burns & H. L. A. Hart (Eds.), *The collected works of Jeremy Bentham*. London: Oxford University Press.
Berg, P. (2004). Asilomar and recombinant DNA. *Nobel Prize Newsletter 8*.
Berg, P., Baltimore, D., Boyer, H., Cohen, S., Davis, R., & Hogness, D. (1974). Biohazards of recombinant DNA. *Science, 185*, 3034.
Bijker, W. (1984). The social construction of facts and artifacts: Or how the sociology of science and the sociology of technology micht benefit each other. *Social Studies of Sciences, 14*, 339–441.
Bijker, W., & Law, J. (1992). *Shaping technology/building society*. Cambridge: MIT Press.
Bijker, W., Thomas, P., & Pinch, T. (Eds.). (1987). *The social construction of technological systems: New directions in the sociology and history of technology*. Cambridge, MA: MIT Press.
Bird, S. (1998). Educational forum: Stimulating a sense of responsibility. *Science and Engineering Ethics, 4*(2), 213–214.
Bleukx, A., & Tampère, C. (2004). *Zelfverklarende wegen: uitbreiding van de verkeersveiligheidstoolbox ontwikkeld aan de K.U. Leuven*. Leuven: KU Leuven.
Bossel, H. (2000). Policy assessment and simulation of actor orientation for sustainable development. *Ecological Economics, 35*(3), 337–355.
Bostrom, N. (2005). Transhumanist values. *Review of Contemporary Philosophy, 4*(5), 1–11.

K. Waelbers, *Doing Good with Technologies: Taking Responsibility for the Social Role of Emerging Technologies*, Philosophy of Engineering and Technology 4,
DOI 10.1007/978-94-007-1640-7, © Springer Science+Business Media B.V. 2011

Bournemouth University. (2008). Animation aids psychology in 'Second Life' experiment. *Science Daily*, March 21.
Briggle, A. (2005). Double effect and dual use. In C. Mitcham (Ed.), *Encyclopeadia for science, technology and ethics* (Vol. 2, pp. 543–546). Farmington Hills: Thompson.
Briggle, A. (2008). Real friends: How the internet can foster friendship. *Ethics and Information Technology, 10*(1), 71–79.
Burgh, M. (1997). Gevolgen van voorspellend erfelijkheidsonderzoek; Alles beter dan onzekerheid. Interview met medisch psycholoog Tibben. *NRC Handelsblad*, 22 Februari 1997.
Callon, M., & Latour, B. (1992). Don't throw the baby out with the bath school! A reply to Collins and Yearley. In A. Pickering (Ed.), *Science as practice and culture* (pp. 343–368). Chigaco: Chicago University Press.
Coeckelbergh, M. (2006). Regulation or responsibility? Autonomy, moral imagination and engineering. *Science, Technology & Human Values, 31*(3), 237–260.
Coeckelbergh, M. (2007). *Imagination and principles: An essay on the role of imagination in moral reasoning*. Basingstoke/New York: Palgrave Macmillan.
Cohen, S., Chang, A., Boyer, H., & Helling, R. (1973). Contruction of biologically functional bacterial plasmids in vitro. *Proceedings of the National Academy of Sciences, 70*(3), 3240–3244.
Collingridge, D. (1980). *The social control of technology*. London: Frances Printer.
Collins, H., & Kusch, M. (1998). *The shape of actions: What humans and machines can do*. Cambridge: MIT Press.
Collins, H., & Yearley, S. (1992). Epistemological chicken. In A. Pickering (Ed.), *Science as practice and culture* (pp. 301–326). Chicago: University of Chicago Press.
Cooper, T. (1987). Hierarchy, virtue, and the practice of public administration: A perspective for normative ethics. *Public Administration Review, 4*, 320–328.
Cooper, R., & Spaight, T. (2007). *Alter Ego: Avatars and their creators*. London: Chris Boot.
Davidse, R. (2007). *Assisting the older driver: Intersection design and in-car devices to improve the safety of the older driver*. Leidschendam: Stichting Wetenschappelijk Onderzoek Verkeersveiligheid (SWOV).
Davidson, D. (1963). Actions, reasons, and causes. *Philosophy, 60*, 685–700.
De Wit, O. (1998). *Telefonie in Nederland: 1877–1940*. Rotterdam: Otto Cramwinckel Uitgever.
Eenink, R. (2009). *Verkeersveiligheidseffecten van Anti-Ongevalsystemen (No. R-2009-11)*. Leidschendam: Stichting Wetenschappelijk Onderzoek Verkeersveiligheid (SWOV).
Ellul, J. (1964). *The technological society* (trans: Merton, R.). New York: Vintage Books.
Ellul, J. (1989). The search for ethics in a technicist society. In F. Ferre & C. Mitcham (Eds.), *Research in philosophy and technology* (Ethics and technology, Vol. 9). London: Jai Press Inc.
European Commission. (2010). Intelligent car initiative, i2010. Retrieved 03 Jan 2010, from http://ec.europa.eu/information_society/activities/intelligentcar/index_en.htm.
Florman, S. (1976). *The existential pleasures of engineering*. New York: St. Martin's Press.
Fogg, B. (2003). *Persuasive technology: Using computers to change what we think and do*. San Francisco: Morgan Kaufmann.
Foucault, M. (1975). *Surveiller et punir, naissance de la prison*. Paris: Gallimard.
Geen, R. (1995). *Human motivation: A social psychological approach*. Pacific Grove: Brooks/Cole Publishing Company.
Gehlen, A. (1961). *Anthropologische forschung*. Reinbek: Rowohlt.
Gietelink, O., Ploeg, J., De Schutter, B., & Verhaegen, M. (2004). *Testing advanced driver assistance systems for fault management with the VEHIL test facility* (No. 04–007). Delft: Delft University of Technology.
Grint, K., & Woolgar, S. (1992). Computers, guns, and roses: What's social about being shot? *Science, Technology, & Human Values, 17*(3), 366–380.
Hara, M. (2004, September). Safety first: VDIM puts Toyota at the head of the safety technology pack in Japan. *Automotive Industries*.
Harbers, H. (1995). Van mensen en dingen, bespreking van Bruno Latour, Wij zijn nooit modern geweest. *Krisis, tijdschrift voor filosofie, 15*(1), 6–15.

Heidegger, M. (1962). Die Frage nach der Technik. In *Die Technik und die Kehre*. Pfullingen: Gunther Neske.
Hendrickxs, L., & Schoot-Uiterkamp, A. (2006). Technology and behavior: The case of passenger transport. In P. Verbeek & A. Slob (Eds.), *User behavior and technology development: Shaping sustainable relations between consumers and technologies*. Dordrecht: Springer.
Hoetink, A. (2003a). *Advanced Cruise Control en verkeersveiligheid* (No. R-2003-24). Leidschendam: Stichting Wetenschappelijk Onderzoek Verkeersveiligheid (SWOV).
Hoetink, A. (2003b). *Advanced cruise control in the Netherlands; A critical review* (paper nr. 4082). Based on SWOV report R-2003-4. *Solutions for today... and tomorrow. Proceedings of the 10th world congress and exhibition on intelligent transportation systems and services ITS*. Brussels: ERTICO –ITS.
Hoeyer, K. (2006). Ethics wars: Reflections on the antagonism between bioethicists and social science observers of biomedicine. *Human Studies, 29*, 203–227.
Ihde, D. (1993). *Postphenomenology*. Evanston: Northwestern University Press.
Ihde, D., & Selinger, E. (2003). *Chasing technoscience: Matrix for materiality*. Bloomington: Indiana University Press.
Illich, I. (1973). *Naar een nieuwe levensstijl*. Baarn: Het Wereldvenster.
Jager, W. (2007). Simulating human behavior for understanding and managing environmental resourse use. *Journal of Social Issues, 63*(1), 97–116.
Jana Chan, M. (2007). Identity in a virtual world: Interview with photographer Robbie Cooper. *CNN*, 14 June 2007.
Jaspers, K. (1957). *Vom Ursprung und Ziel der Geschichte*. Frankfurt: Fischer Bucherei.
Jaspers, K. (1962). *Die Atombombe und die Zukunft des Menschen*. Munchen: Piper.
Jelsma, J. (2000). *Design behavior steering technology*. Paper presented at the summer academy on Technology Studies 2000, IFZ, Graz.
Johnson, M. (1993). *Moral imagination*. Chicago: University of Chicago Press.
Jonas, H. (1984). *The imperative of responsibility: In search of an ethics for the technological age*. Chicago: University of Chicago Press.
Jones, S. (2006). *Against technology: From Luddites to neo-Luddism*. New York: Routledge.
Kant, I. (1785/1993). *Grounding for the metaphysics of morals* (trans: Ellington, J.). Indianapolis: Hackett.
Khong, L. (2003). Actants and enframing: Heidegger and Latour on technology. *Studies in History and Philosophy of Science, 34*, 693–704.
Kline, R., & Pinch, T. (1996). Users as agents of technological change: The social construction of the automobile in the rural United States. *Technology and Culture, 37*, 763–795.
Lash, S., & Lury, C. (2007). *Global culture industry: The mediation of things*. Cambridge: Polity Press.
Latour, B. (1987a). Hoe "de Heerser" te schrijven voor zowel machinaties als machines. *Krisis, 7*(26), 42–66.
Latour, B. (1987b). *Science in action*. Cambridge, MA: Harvard University Press.
Latour, B. (1988). How to write "The Prince" for machines. In B. Eliot (Ed.), *Technology and social change* (pp. 20–43). Edinburgh: Edinburgh University Press.
Latour, B. (1991). *We have never been modern*. Cambridge, MA: Harvard University Press.
Latour, B. (1992). Where are the missing masses? The sociology of the new mundane artefacts. In *Shaping technology, building society*. Cambridge: MIT Press.
Latour, B. (1994). On technical mediation: Philosophy, sociology, genealogy. *Common Knowledge, 94*(4), 29–64.
Latour, B. (1995a). De antropologisering van het wereldbeeld: een persoonlijk verslag. *Krisis, 58*, 29–37.
Latour, B. (1995b). A door must be either open or shut: A little philosophy of techniques (trans: Cussins, C.). In A. Feenberg & A. Hannaway (Eds.), *Technology, and the politics of knowledge* (pp. 272–281). Bloomington: Indiana University Press.
Latour, B. (1996). *Aramis or the love of technology*. Cambridge, MA: Harvard University Press.

Latour, B. (1999a). On recalling ANT. In J. Law & J. Hassard (Eds.), *Actor network theory and after*. Oxford: Blackwell Publishers.

Latour, B. (1999b). *Pandora's hope: Essays on the reality of science studies*. Cambridge, MA: Harvard University Press.

Latour, B. (2002). Morality and technology: The end of the means. *Theory, Culture & Society, 19*(5/6), 247–260.

Latour, B. (2005a). *Reassembling the social: An introduction to actor-network-theory*. Oxford: Oxford University Press.

Latour, B. (2005b). Van Realpolitik naar Dingpolitik. *Krisis, 2*, 40–61.

Latour, B. (2008). *Coming out as a philosopher: Acceptance speech for the third Siegfried Unseld Prize*. Acceptance speech for the third Siegfried Unseld Prize.

Lenk, H. (1993). Uber Verantwortungsbegriffe und das Verantwortungsproblem in der Ethik. In H. Lenk & G. Ropohl (Eds.), *Technik und Ethik* (2nd ed., pp. 112–148). Stuttgart: Reclam.

Levitt, T. (1983). The globalisation of markets. *Harvard Business Review, 3*, 92–102.

Lindeman, G. (2009). *From experimental interaction to the brain as the epistemic object of neurobiology*. Paper presented at the Artificial by Nature: IVth International Plessner Conference, Rotterdam.

MacIntyre, A. (1985). *After virtue* (2nd ed.). London: Duckworth.

MacIntyre, A. (1988). *Whose justice, which rationality?* London: Duckworth.

MacIntyre, A. (1990). *Three rival versions of moral enquiry: Encyclopedia, genealogy and tradition*. Notre Dame: University of Notre Dame Press.

MacIntyre, A. (1998). Politics, philosophy and the common good. In K. Knight (Ed.), *The MacIntyre reader*. Notre Dame: University of Notre Dame Press.

MacIntyre, A. (1999). *Dependent rational animals: Why human beings need the virtues*. London: Duckworth.

MacIntyre, A. (2006a). *Ethics and politics* (Selected essays, Vol. 2). Cambridge: Cambridge University Press.

MacIntyre, A. (2006b). What is a human body? In A. MacIntyre (Ed.), *The task of philosophy* (Selected essays, Vol. 1). Cambridge: Cambridge University Press.

MacIntyre, A. (2007). Social structures and their threats to moral agency. In A. MacIntyre (Ed.), *Ethics and politics* (Selected Essays, Vol. 2). Cambridge: Cambridge University Press.

MacIntyre, A. (2008). What more needs to be said? A beginning, although only a beginning. In K. Knight & P. Blackledge (Eds.), *Revolutionary Aristotelianism: Ethics, resistance and utopia*. Stuttgart: Lucius & Lucius.

MacIntyre, A. (2009a). *God, philosophy, universities: A selective history of the catholic philosophical tradition*. Plymouth: Rowman & Littlefield.

MacIntyre, A. (2009b). Open session for questions at the Alasdair MacIntyre Conference: On having survived the moral philosophies of the twentieth century, *Alasdair MacIntyre Conference: On having survived the moral philosophies of the twentieth century*. Dublin: University College Dublin.

Marcuse, H. (1964). *One-dimensional man: Studies in the ideology of advanced industrial society*. Boston: Beacon.

Martin, M. (2006). Moral creativity in science and engineering. *Science and Engineering Ethics, 12*(3), 421–433.

Marx, K. (1963/1847). *The poverty of philosophy*. New York: International Publishers.

Mazzae, E., & Forkenbrock, G. (2003). *VRTC crash avoidance research: Light vehicle ABS research*. Washington DC: NHTSA.

Meadows, D., Meadows, H., et al. (1972). *The limits to growth: A report for the Club of Rome's project on the predicament of mankind*. New York: Universe Books.

Meijers, A. (2000). *The empirical turn in the philosophy of technology*. Greenwick, CT: JAI Press.

Mele, A. (2006). Action. In D. Borchert (Ed.), *Encyclopedia of philosophy* (2nd ed., Vol. 1, pp. 14–22). Detroit: Macmillan Reference USA.

Merton, R. (1973). *The sociology of science*. Chicago: University of Chicago Press.

Mills, C. (1959). *The sociological imagination*. London: Oxford University Press.

Misa, T. (1988). How machines make history and how historians (and others) help them to do so. *Science, Technology, & Human Values, 13*(3/4), 308–331.

Mitcham, C. (1985). Langdon Winner on Jacque Ellul: An introduction to alternative political critiques of technology. In S. H. Cutcliffe (Ed.), *Contemporary critiques of technology* (Vol. 3, pp. 91–114). Bethlehem: Technology Studies Resource Center.

Mitcham, C. (1994). *Thinking through technology: The path between engineering and philosophy*. Chigaco: The University of Chicago Press.

Mitcham, C. (1997). Engineering design research and social responsibility. In K. Shrader-Frechette & L. Westra (Eds.), *Technology and values*. Boston Way: Rowman & Littlefield Publishers.

Mitcham, C. (2003). Co-responsibility for research integrity. *Science and Engineering Ethics, 9*(2), 273–290.

Mitcham, C., & Waelbers, K. (2009). Technology and ethics: Overview. In J. Berg Olsen, S. Pedersen, & V. Hendricks (Eds.), *A companion to the philosophy of technology* (pp. 367–383). West Sussex: Wiley Blackwell.

More, T. (2005). *Utopia*. London: Elibron Classics Series.

Mumford, L. (1934). *Technics and civilization*. New York: Harcourt, Brace and World.

Murphy, M. C. (2003). MacIntyre's political philosophy. In M. C. Murphy (Ed.), *Alasdair MacIntyre*. Cambridge: Cambridge University Press.

Nicholas, O. (2001). *Medieval children*. New Haven: Yale University Press.

Noorman, M. (2008). Limits to the autonomy of agents. In A. Briggle, K. Waelbers, & P. Brey (Eds.), *Current issues in computing and philosophy*. Amsterdam: IOS Press.

Nussbaum, M. (1986). *The fragility of goodness: Luck and ethics in Greek tragedy and philosophy*. Cambridge: Cambridge University Press.

Nussbaum, M. (1995). *Poetic justice: The literary imagination and public life*. Boston, MA: Beacon.

Nussbaum, M. (1997). *Cultivating humanity: A classical defense of reform in liberal education*. Cambridge, MA: Harvard University Press.

Oei, H. (2003). *Ervaringen met Advanced Cruise Control in een korte praktijkproef (No. D-2003-4)*. Leidschendam: Stichting Wetenschappelijk Onderzoek Verkeersveiligheid (SWOV).

Pence, G. (1991). Virtue theory. In P. Singer (Ed.), *Blackwell companions to philosophy: A companion to ethics* (pp. 249–259). Oxford: Blackwell Publishers.

Pew, R., & Mavor, A. (1997). *Representing human behavior in military simulations: Interim report*. Washington, DC: National Research Council.

Pickering, A. (1995). *The mangle of practice*. Chicago: The University of Chicago Press.

Pinkard, T. (2003). MacIntyre's critique of modernity. In M. C. Murphy (Ed.), *Alasdair MacIntyre*. Cambridge: Cambridge University Press.

Postman, N. (1992). *Technopoly*. New York: Alfred A. Knopf, Inc.

Pritchard, M. (2001). Responsible engineering: The importance of character and imagination. *Science and Engineering Ethics, 7*(3), 391–402.

Rawls, J. (1999). *A theory of justice* (Revised edition). Cambridge, MA: Harvard University Press.

Richardson, H. (1999). Institutionally divided moral responsibility. In E. Frankel-Paul, F. Miller, & J. Paul (Eds.), *Responsibility* (pp. 218–249). New York: Cambridge University Press.

Rip, A., Misa, T., & Schot, J. (1995). *Managing technology in society: The approach of constructive technology assessment*. London: Pinter Publishers.

Roeser, S. (2005). Defending moral intuition. In R. van Woudenberg, S. Roeser, & R. Rood (Eds.), *Basic belief and basic knowledge: Papers in epistemology* (pp. 231–250). Frankfurt: Ontos Verslag.

Ropohl, G. (1993). Neue Wege, die Technik zu verantworten. In H. Lenk & G. Ropohl (Eds.), *Technik und Ethik* (2nd ed., Vol. 8395, pp. 149–176). Stuttgart: Reclam.

Rosen, P. (2002). *Framing production: Technology, culture and the change in british bicycle industry*. Cambridge, MA: MIT Press.

Schot, J., & Rip, A. (1997). The past and future of constructive technology assessment. *Technological Forecasting and Social Change, 54*, 251–268.

Sismondo, S. (2004). *Science and technology studies*. Oxford: Blackwell Publishing.

Smit, W., & van Oost, E. (1999). *De wederzijdse beinvloeding van technologie en maatschappij: een technology-assessment benadering*. Bussum: Uitgeverij Coutinho.

Smith, A. (1776). *An inquiry into the nature and causes of the wealth of nations*. London: W. Strahan and T. Cadell.

Smith, A. (2003). Do you believe in ethics? Latour and Ihde in the trenches of the science wars (or: Watch out, Latour, Ihde's got a gun). In D. Ihde & E. Selinger (Eds.), *Chasing technoscience: Matrix for materiality* (pp. 182–194). Bloomington: Indiana University Press.

Smits, R. (2000). *Innovatie in de Universiteit: rede uitgesproken bij de aanvaarding van het ambt van hoogleraar 'Technologie en Innovatie'*. Utrecht: Universiteit Utrecht.

Smits, R., & Leyten, A. (1991). *Technology assessment: waakhond of speurhond: naar een integraal technologiebeleid*. Zeist: Kerckebosch b.v.

Staudenmaier, S. J. (1985). *Technology's storytellers, reweaving the human fabric*. Cambridge, MA: MIT Press.

Stemerding, D., & Swierstra, T. (2006). How might interactive scenariostudies help us to think about the normative implications of genomics and predictive medicine. In A. de Bouvet, P. Boitte, & G. Aiguier (Eds.), *Questions éthiques en médicine prédictive* (pp. 81–88). Paris: John Libbey Eurotext.

Stoner, J., Evans, D., & Mcgehee, D. (1997). *Development of vehicle simulation capability*. Berkeley: California Partners for Advanced Transit and Highways, University of California.

Sundström, P. (1998). Interpreting the notion that technology is value-neutral. *Medicine, Health Care and Philosophy, 1*, 41–45.

Swierstra, T. (1992). Latour de force. *Kennis en methode, 1*, 21–38.

Swierstra, T. (1995). Een koud bad. *Krisis, tijdschrift voor filosofie, 15*(1), 25–28.

Swierstra, T. (1999). Moeten artefacten moreel gerehabiliteerd? *K&M – tijdschrift voor empirische filosofie, 4*, 317–326.

Swierstra, T. (2000). Bruno Latour. In M. Doorman & H. Pott (Eds.), *Filosofen van deze tijd* (pp. 357–372). Amsterdam: Bert Bakker.

Swierstra, T. (2010). Het huwelijk tussen techniek en moraal. In M. Huijer & M. Smits (Eds.), *Moralicide. Mens, techniek en symbolische orde. [Jaarboek Civis Mundi i.s.m. Rathenau Instituut]* (pp. 17–35). Rotterdam: Lemniscaat.

Swierstra, T., Stemerding, D., & Boenink, M. (2009). Exploring techno-moral change: the case of the obesity pill. *Humanities, Social Sciences and Low 3*(2), 119–138.

Swierstra, T., & Jelsma, J. (2005). Trapped in the duality of structure: An STS approach to engineering ethics. In H. Harbers (Ed.), *Inside the politics of technology*. Amsterdam: Amsterdam University Press.

Swierstra, T., & Jelsma, J. (2006). Responsibility without moralism in technoscienfific design practice. *Science, Technology and Human Values, 31*(3), 309–332.

Swierstra, T., & Rip, A. (2007). Nano-ethics and Nest-ethics: Patterns of moral argumentation about new and emerging science and technology. *Nanoethics, 1*(1), 3–20.

Swierstra, T., Stemerding, D., & Boenink, M. (2009). Exploring techno-moral change: The case of the obesity pill. In P. Sollie & M. Düwell (Eds.), *Evaluating new technologies: Methodological problems for the ethical assessment of technology developments*. Dordrecht: Springer.

SWOV. (2008a). *SWOV-Factsheet: Inhoud en evaluatie van verkeerseducatieprogramma's*. Leidschendam: Stichting Wetenschappelijk Onderzoek Verkeersveiligheid.

SWOV. (2008b). *SWOV fact-sheet: Advanced cruise control*. Leidschendam: Stichting Wetenschappelijk onderzoek verkeersveiligheid (SWOV).

SWOV. (2010). *SWOV-Factsheet. Herkennen van gevaren in het verkeer*. Leidschendam: SWOV.

Tabula Rasa. (2004). *Positieve wegen tot gedragsbeïnvloeding: motieven en interventies voor veilig verkeersgedrag*. Den Haag: Tabula Rasa.

Taylor, C. (1989). *Sources of the self: The making of the modern identity*. Cambridge: Cambridge University Press.

Tenner, E. (1997). *Why things bite back: Technology and the revenge of unintended consequences*. New York: Vintage Books.

Tenner, E. (2004). *Our own devices: How technology remakes humanity*. New York: Vintage Books.

Thaler, R., & Sunstein, C. (2008). *Nudge: Improving decisions about health, wealth and happiness*. London: Penguin Books.

Thomson, D. (1980). Moral responsibility of public officials: The problem of many hands. *The American Political Science Review, 74*(4), 905–916.
Tibben, A. (2000). *Van vrees naar hoop: erfelijke neurodegeneratieve ziekten opnieuw bezien. Rede uitgesproken bij de openlijke aanvaarding van het ambt van bijzonder hoogleraar aan de Faculeit der Geneeskunde.* Leiden: University of Leiden.
Tibben, A., Timman, R., Bannink, E., & Duivenvoorden, H. (1997). Three-year follow-up after presymptomatic testing for Huntington's disease in tested individuals and partners. *Health Psychology, 16*, 20–35.
Tonnon, H., & van der Jagt, D. (2003). Information about safe use of Advanced Cruise Control in the car is necessary. *Leidschendam Press release SWOV.*
Treece, J. (1997). Toyota offers radar on Japan's Celsior. *Automitive News 8.*
Turner, D., & Hartzell, L. (2004). The lack of clarity in the precautionary principle. *Environmental Values, 13*, 449–460.
van Dijk, P. (2000). *Anthropology in the age of technology: The philosophical contribution of Gunther Anders.* Amsterdam: Rodopi.
Verbeek, P. (2005a). *The morality of design: Some notes on the moral agency of artefacts, users and designers.* Paper presented at the SPT 2005 Conference 'Technology and designing', Delft, Delft University of Technology.
Verbeek, P. (2005b). *What things do – Philosophical reflections on technology, agency and design.* Penn State: University Press.
Verbeek, P. (2006). Materializing morality: Design ethics and technological mediation. *Science, Technology and Human Values, 31*(3), 361–380.
Verbeek, P. (2008a). Cultivating humanity: Towards a non-humanist ethics of technology. In J.-K. B. Olsen, E. Selinger, & S. Riis (Eds.), *New waves in philosophy of technology* (pp. 241–266). Hampshire: Palgrave MacMillan.
Verbeek, P. (2008b). Obstetic ultrasound and the technological mediation of morality: A post phenomenological analysis. *Human Studies, 31*(1), 11–26.
Verbeek, P. (2009). *Oratie: Techniek en de grens van de mens: over techniek en de menselijke natuur.* Enschede: University of Twente.
Waelbers, K. (2002). Blame it on the technology. *NVBE Newsletter 2.*
Waelbers, K. (2009a). From assigning to designing technological agency. *Human Studies, 32*(2), 241–250.
Waelbers, K. (2009b). Technological delegation: Responsibility for the unintended. *Journal for Science and Engineering Ethics, 15*, 51–68.
Waelbers, K. (2009c). Slapende Politiemannen: Disciplinerende technologieën in het verkeer. In Frits Bolkestein, Jeroen van den Hoven, Ibo van de Poel en Ilse Oosterlaken (Eds.), *Politiek der Dingen: Over technische artefacten en hun morele en politieke lading.* Budel: Uitgeverij DAMON.
Waelbers, K. & Briggle, A. (2010) Technology, the good life, and liberalism: Some reflections on two principles of neutrality. *Techné: Research in Philosophy and Technology, 14*(3), 176–193.
Waelbers, K., Stafleu, F., & Brom, F. (2004). Not all animals are equal: Differences in moral foundations for the Dutch veterinary policy on livestock in nature reservations. *Journal of Agricultural and Environmental Ethics, 17*, 497–515.
Weegink, R. J. (1996). *Basisonderzoek elektriciteitsverbruik kleinverbruikers BEK'95.* Arnhem: EnergieNed.
Weinberg, A. M. (1966). Can technology replace social engineering? *Bulletin of the Atomic Scientists, 12*(10), 4–8.
Winner, L. (1980). Do artifacts have politics. *Daedalus, 109*(1), 121–136.
Winner, L. (1988). The whale and the reactor: A personal memoir. *Journal of American Culture, 3*(448), 446–455.
Woolgar, S., & Cooper, G. (1999). Do artefacts have ambivalence? *Social Studies of Science, 29*(3), 433–449.

Index